Emergent Chemical Evolution

THE ORIGIN OF LIFE SOLVED

EDUARDO TRISAPIENT HERNANDEZ

Copyright © 2018 Eduardo Trisapient Hernandez

All rights reserved.

ISBN: 198539846X
ISBN-13: 978-1985398467

DEDICATION

With Love to my wife Maria Fernanda, and our children Eresvey, Eduardo, Ethan, and Elyza. Without whom, I would not have found myself in the right circumstances and situations to discover Emergent Chemical Evolution.

CONTENTS

	Acknowledgments	i
1	Introduction	1

Section I Local Conditions of Prebiotic Earth

2	The Relentless Currents	6
3	Carbon Dioxide in the Atmosphere	9
4	The Abundance of Amino Acids	13
5	Iron Dissolved in the Oceans	15

Section II Difficulties of Other Hypothesis

6	List of Erroneous Assumptions	19
7	The Conditions of the Prebiotic Oceans	20
8	Self-replicating Molecule Initiated Life	22
9	Lipid Cell Membrane Needed to Initiate Life	24
10	Viruses Originated from Living Cells	27
11	Homeostasis Was Needed to Initiate Life	29
12	Life Occurred Randomly	31

Section III Emergent Chemical Evolution

13	Putting It All Together	33
14	Structure Not Energy Is Most Important	35
15	The World As a System	40
16	The Currents of Life	43

17	The Chaos of ECE	51
18	Inevitability of Life	54
19	From Kinetic Energy to Chemical Energy	73
20	The First Life Form	77
21	Dumb RNA	82
22	The Rise of Ribosomes	88
23	I'm Right Handed, Why Are My Aminos Left?	94
24	The Indestructible Stage of Life	98
25	Multiple Strains, Multiple Lineages	106
26	A Precellular Ecosystem is Formed	112
27	The Rise of Eukaryotes	118
28	Tying the Emergence of Life With the Emergence of the Universe	125
	Review of Evidence for ECE	127
	About the Author	128
	References	129
	Photo and Illustration Credits	132
	Cover Designed	133

ACKNOWLEDGMENTS

I would like to acknowledge the tremendous work of the National Institute of Health for the creation and maintenance of their publicly available Internet database. I would also like to acknowledge the long standing tradition of excellence of the Journal of Biological Chemistry and for providing the free access of their old papers online. I also would not have been able to complete my work without the availability of Cornell University Library's arXiv.org online open access.

1 INTRODUCTION

Back in 2004 in order to complete my Chemistry Degree, I decided to take an Introduction to Philosophy class. It was here that the professor informed us that one of life's biggest questions, exactly how life had started on Earth, was not yet answered! I was so taken aback by this revelation that I nearly toppled over backwards in my chair. For nearly two decades I was under the mistaken impression that we knew exactly how it all had happened. The reason I was under such an impression was because back in the summer of 1987 I had read the Scientific American book called *Molecules to Living Cells*. By the end of which, I remember, being very convinced that we already knew exactly how life had started; but then again, I was only seventeen and, I guess, very impressionable.

Well that same day in 2004 when I returned home from my philosophy class, I found Molecules to Living Cells and began rereading it. By the time I had read the first quarter of the introduction, I realized that- yes- I was a very impressionable seventeen year-old. Back in 1987 I was only five years an atheist, and just beginning to hone my critical thinking skills. Impressionable or not, the articles in that book are profound and very convincing. What else could have been expected from an impressionable seventeen year old, when they were written by some of the most brilliant scientists of the twentieth century. So I thought, consoling myself, with that kind of mental prowess behind these articles anyone would have been convinced; especially, when it fits their world view.

Shortly after reading the book for the first time, my world view and belief- that we knew exactly how life had started- was further solidified when I became a junior biomedical investigator in the Institute of Animal Behavior at Rutger's University in Newark, New Jersey during the summer of 1989. There, under the guidance of Dr. Barry Komizaruk, I learned how to conduct double-blind experiments and do proper research using scientific journals. While working in that laboratory, I assisted Dr. Komizaruk and his colleagues with their experiments in neuro-anatomy, isotope assays of proteins, and even performed surgery on live rats as controls for their experiments. After seeing the scientific method in action, up close and personal, I was convinced that there was nothing- with its proper use- it could not solve. This belief was also continuously reinforced by the scientific method's ever increasing body of published successes, and my witnessing of their application to an array of astounding modern technologies.

It was, precisely, because I was so aware of these successes that I was later so taken aback by my philosophy professor's statement. Until that moment, I was content in my world view and had held it as being irrefutable. That was also precisely the moment that I realized what my new goal in life would be. That new goal was to develop the theory that would precisely explain how life began on Earth. Since that moment, and for over a decade, I have incessantly contemplated and researched this very question until I came up with its solution! This book is the explanation of that solution!

I had researched this question so incessantly, that I alienated all my friends and relatives; almost my wife and kids too. I had spent nearly every waking moment reading through all my books on: organic chemistry, biochemistry, molecular cell biology, and genetics (among others); and when I was not reading, I was on-line. On-line researching the questions that I produced during my time contemplating my theory. Which was most of the time that I was not either reading or on-line. This incessant contemplation was also what made me more forgetful, absent minded, and not present in my personal life; making it very empty. In spite of all that, and with my wife and kids' continued love and support, I was able to achieve my goal!

The elucidation of that goal is the purpose of this book. The goal- that I worked for over a decade to achieve- was to produce a theory that would explain exactly how life emerged on Earth. Well, after so many years of dedication and sacrifice, not only did I produce a theory that works and has published evidence supporting it, I also discovered an emergent mechanism that is the central theme of my theory. A mechanism that arose from the undeniable and ever present forces of nature. A mechanism that was subject to evolutionary pressures and natural selection. A mechanism that- once set in motion- would inevitably result in the emergence of life on Earth or anywhere else these same conditions would converge.

This mechanism resulted in very specific conclusions about what its output would have been and what traces would have been left behind. Every time I came up with a specific trace of what this mechanism would have left behind; I researched it on-line. Every time I did so, I found evidence for it and not just any evidence- published peer-reviewed evidence. Published articles written by the most respected scientists in their respective fields. So there can be no question as to the validity of these articles; therefore every referenced article in this book should be seen as another validation of the predictive ability of my theory. That being the reason why- in this book- I have taken the bold step of calling my repeatedly confirmed hypothesis a theory.

In section one, Local Conditions of Prebiotic Earth, I elaborate first on the little known fact that, until recently, there was no consensus among geologist on what were the exact conditions on the Prebiotic Earth. I then use articles published more recently to show how my assertions on the conditions of Prebiotic Earth are supported. Next, I used modern observations of the forces of nature to help elucidate the conditions of Prebiotic Earth. And thus by showing how my assertions are supported I set the stage for explaining how my theory functioned.

In section two, Difficulties of Other Hypothesis, I list a series of assumptions that origin of life investigators have been using for years. I then proceed to break down every assumption, and show why these do not clarify how life arose on Earth. I then pose questions about life that naturally arise when these assumptions are challenged. These questions serve to help us better understand what life actually is; therefore, setting the stage where Emergent Chemical Evolution can answer these questions and fill in the gaps.

Finally in the section, Emergent Chemical Evolution, I give a detailed description of my theory. In this section I have one chapter that mathematically shows the output of its central emergent mechanism. In this section I also present in very clear materialistic steps an explanation of how chemicals coalesced and how their products evolved into becoming life. Then -for every materialistic step I describe- I present a published article by an imminent scientist in that field as evidence. Again and again at every step presenting evidence after evidence of why my theory, Emergent Chemical Evolution, is the best solution for the emergence of Life.

Please Review this book on Amazon.com

Connect with me on:

www.facebook.com/EmergentChemicalEvolution/

@Trisapient

Section I
Local Conditions of Prebiotic Earth

2 THE RELENTLESS CURRENTS

I once saw a television program about the organisms that live in the oceans. That program reported that the currents of the oceans bring nutrients from other more fruitful parts of the oceans and provides it to life that resides in more barren parts of the oceans. That program also stated that the currents never stop. They are constantly shifting material throughout the oceans. I was also very impressed by how far the currents moved all that material. At the Earth's polar caps the oceans' water is cooled down and then sinks to the sea-floor. Where it would be once again taken by the currents to the equatorial regions delivering oxygen to the sea life there. Where it again loses oxygen, heats up, and rises to the surface of the ocean. Where the water once again is taken by the oceanic currents-all the while regaining oxygen- back to the Polar Regions. And this recycling, the program said, has been going on since the formation of the oceans.

    Now jump a couple of years later to my organic chemistry lab. Where at that instant, I was reacting two chemical compounds inside a beaker which was on crushed ice. At that moment I was thinking to myself "How did nature provide the cold temperature for these reactions to occur so that life could start?" At that instant, I remembered the program that I had watched and what It had taught me about the oceanic currents going to the Polar Regions. If currents have been running since the formation of the oceans, then the prebiotic Earth could have easily provided cold, as well as hot, conditions for any chemicals contained in its oceans.

When I got home after lab, I did a little more research on oceanic currents. I discovered that oceanographers had created a computer simulation of a planet with only water on its surface. When they ran the simulation they discovered that the rotation of the planet, by itself, would create currents in its oceans. These currents would form without the need of heat from this planet's sun. They formed solely from the gravity and rotation of the planet. Also because there were no land masses, the currents circulated in bands around the planet. What I also then realized was that- as soon as the oceans formed on the early earth- currents would have immediately formed as well; churning up the contents of the prebiotic oceans. Exposing that content to different conditions around the globe.

While I was thinking about that, another memory resurfaced. This time, it was about swimming at the community pool as a teenager. One summer, I do not remember which, the lifeguard formed us into a circle inside the pool while holding hands. He then instructed us to walk in a clockwise direction fallowing the walls of the pool. After only traveling half way around the pool, he instructs us- the tallest and strongest boys- to get out of the pool. Needless to say- we failed to get out and my shorts ended up at the other end of the pool in the attempt. The current was so relentless it would not let me escape. The point of the exercise, of course, was for us not to underestimate the power of water currents. Little did they, nor I, know then that that simple and valuable lesson would lead me to discover how Life began on Earth.

Those vivid memories also brought about other realizations about how life must have begun. One was that the relentless currents had to be involved in the formation of life. Another was that, not only did, the relentless currents expose their contents to different conditions that existed at different parts of the globe at that time, but that they would have endlessly recycled them. These relentless currents pose a major problem for those hypothesis of abiogenesis that require accumulation of material in rock cavities or on mineral surfaces. These relentless currents would have stripped any cavity of its content and ripped away any starting material located on mineral surfaces before they would have had enough time to react effectively. This fact also poses a problem for any hypothesis that requires prolonged exposure to local phenomena- such as volcanic vents; therefore, any hypothesis of abiogenesis would either have to account for these relentless currents, explain how they are insignificant, or said theory must be abandoned.

    These relentless currents were the starting point for my investigations into the origins of life. That the Earth was covered in water at life's beginning- as far as I knew- has never been challenged. So any theory of abiogenesis would have to not only account for them but incorporate those currents. Even if the current was moving at a snails pas, do you know how many hydrogen or oxygen atoms would rub up against any reacting molecules by the passage of just one inch of water? Well let us work it out. A water molecule is approximately 2.75 angstroms long. There are 254 million angstroms in an inch. We then divide 254 million by 2.75 we get 92.36 million. So at the very least, because it is a liquid, we would have 92.36 million water molecules rubbing against the reactants. Everyone of them either wanting to form a hydrogen bond or to bond with an oxygen atom from an hydroxide ion. Any working theory would have to do one of three things. One is to completely shield molecules from the currents which further adds to the problem of explaining how, why, and then when they got free after reacting. Two, show how the currents would not effect the reactants. Or three, show how those currents helped the reactions take place; which is the position I take in this, my, theory.

3 CARBON DIOXIDE IN THE ATMOSPHERE

The amount of carbon dioxide has undoubtedly increased in the last hundred years-thanks to the activities of man. The amazing thing, about this problem, is that the content of carbon dioxide in the atmosphere is just 0.0391 % (as of 10/2012). This amazingly small amount of carbon in the air is enough to cause the acidification of todays oceans. This small acidification of the oceans is still such that it is causing havoc on thousands of coral colonies by bleaching them and keeping millions of plankton from forming complete skeletons. This small amount of carbon dioxide is still large enough to have a major impact on life in the oceans today. Life is impacted despite being self-contained organisms with membranes; impacted despite having mechanisms to maintain their internal pH; impacted despite there being a fully oxygenated atmosphere. Would it not then stand to reason, that a larger amount of carbon dioxide in the atmosphere on the prebiotic Earth would have had a much greater effect on the origin of life? Especially when there was little free oxygen, if at all, then?

Current estimates- based on geological evidence from the Hadean era- suggest that the amount of carbon dioxide in the atmosphere could have been higher than 10 %. The impact of such a large amount of carbon dioxide on the oceans back then is very hard to imagine, especially, when such a small amount has had such a large devastating impact on modern oceans and its life today.

The first thing that we need to consider is the amount of atmospheric pressure that was present back then. Current estimates have it that the atmospheric pressure was between 20 and 480 times greater than it is today. The atmospheric pressure that existed then is important to consider because of its very particular effect on the pH of the oceans. The increased atmospheric pressure of course decreased the pH of the oceans, but that is not the whole story.

Figure 1.
This much bleaching is done by the acidification of the oceans by the miniscule air content of 0.0391% CO2(as of 10/2012). How acid would the oceans be if the CO2 content of the atmosphere were at least 10%?
(Photo courtesy of National Oceanic and Atmospheric Administration)

The pH of the oceans would have been lesser, more acidic, in the water that was closer to its surface. This is the case due to the partial pressure rules of gcarbon dioxide gas present over water which dictates this behavior. This behavior can also be confirmed as evidenced by the acidification of today's oceans. The pH of the ocean is more acidic near the surface and less so the deeper you measure. The bleaching of the coral reefs happens at very low depths, very near the surface, of the oceans. That is where the partial pressure of the carbon dioxide of the atmosphere would have the greatest effect.

But what I consider to be even more important is the drop in acidity the further down you go into the depths of the ocean. The deeper you go into the ocean the less effect the atmospheric carbon dioxide has on the ocean's pH.

I can already hear everyone's objections repeating the same age old litany: "The average pH of the oceans at this time is believed to still have been basic." Currently, there are more that, correctly, state that the oceans were acidic but that its average pH was 6.3. But I am going to say that the average pH of the ocean is as elucidating as knowing the average atmospheric temperature today. In other words can knowing the average atmospheric temperature today tell you anything about the present temperature on the North Pole or in Hawaii? No. Neither does knowing the average atmospheric pressure tell you if there was a hurricane in Florida or a monsoon in Asia. The average of any measurable quantity does not tell you anything about local conditions at any given time; thus, the need for a closer look at the effects of carbon dioxide on the pH of the oceans is still warranted.

Well then, with the Hadean eon carbon dioxide content at or above 10%- the pH of the ocean near the surface would have been very low. Easily below 3.4 pH the calculated pH of a solution with carbon dioxide gas above it at 10 atms. Again, what is most important for us to consider is the gradient of pH that emerges thanks to the high carbon dioxide content of the atmosphere. The deeper, thus farther from the partial pressure, the higher the pH gets. Eventually the pH of the ocean at a deeper depth will be 5 pH. This is the pH of cell organelles called lysosomes where, at this 5 pH, certain proteins become active and breakdown the chemical bonds of the nutrients that that cell ingests. Go deeper and you will come across 7.2 pH, the pH that most cells need their cytoplasm to be at in order for them to function properly or replicate.

For a couple of years I thought that this was all I could say about the matter of pH gradients forming in the prebiotic oceans. But as it turned out- to my utter surprise- I came across another published article with modern corroborating evidence on this very subject. As it turns out Jaroslav Flegr reports that eukaryotic cells set up a pH gradient within their cytoplasms which make proteins function better and make those cells able to become 3 to 4 times larger than prokaryotic cells (Flegr 2009). This article not only provides us with proof that a pH gradient is beneficial to life which is more modern, corroborating, and published evidence in support of my theory. It also shows how my theory once again solves yet another unforeseen problem. My theory will also show how Emergent Chemical Evolution took advantage of the pH gradients that emerged in the oceans, and it will also give a means by which life would have evolved to mimic this very characteristic.

As you can begin to see, the ocean was not a static alkaline mass of water as some origin of life investigators portray it to have been. So let us review: 1. the gravity and rotation of the earth created relentless currents that carry material throughout the oceans. 2. Those relentless currents would have endlessly cycled that material throughout the different pH layers of the oceans. Next, we need to determine what kind of material was endlessly cycled within those oceans.

4 THE ABUNDANCE OF AMINO ACIDS

When I was formulating this theory, I thought I would need to convince the reader that meteorites delivered most of the amino acids needed to initiate Emergent Chemical Evolution (ECE). But thanks to Z. Martins and others I do not have to try so hard anymore; because back in 2007 they wrote an article on this very subject. In which they claim that the amino acids delivered by chondrites would have played a part in starting life on earth (Martins and others 2007). This is one of the central claims of Emergent Chemical Evolution.

    The necessity of some scientists to prove how amino acids were produced on Earth from scratch seems to be totally unwarranted. Especially when, for over thirty years, it has been commonly known that chondrites contain, since before the formation of the solar system, over seventy different types of amino acids. And there is overwhelming evidence that there was a period of heavy bombardment just prior to the emergence of life. I propose that these meteorites were the primary source of the amino acids that were necessary for the emergence of life and that it was also unnecessary for the Earth to produce any amino acids for that purpose.

    Well that was going to be my argument anyway. But now I feel that the authors of the previous article have very eloquently proven my point. But it does not end there. Another article published on August 10, 2012 in

Science magazine goes even further than that. Alexander et al proved that chondrites not only contributed amino acids to earth but also delivered all volatile material: water, hydrogen, carbon, and nitrogen (Alexander et al 2012).

One of the problems faced by geologist was explaining how the Earth got its water and other volatile material after it had melted soon after its formation. Once the Earth had amassed itself from the debris of the solar nebula at some point it reached a critical mass. When that critical mass was reached something spectacular occurred, between its own gravitational pull and the amount of nuclear material contained within it, the entire Earth melted. That means that every inch of the entire planet would have been above 1300°F. At that temperature all volatile material like water, carbon, hydrogen, all gases would have been burnt and irradiated out into space. Alexander et all, I believe, have provided compelling evidence of not only the source of the oceans but also of its contents as well.

To clarify the significance of these two articles for the Theory of Emergent Chemical Evolution I need to add the fallowing statements: First, these two articles make it abundantly clear that the prebiotic oceans were full of amino acids. Second, that without the massive bombardment there would have been no starting material for the Earth to cook up amino acids in the first place. Therefore, there is no need to investigate or to recreate chemical mechanisms that would help explain how amino acids formed on Earth. The amino acids that were needed, for Emergent Chemical Evolution to occur, were not synthesized on Earth but were simply brought here by meteorites during the heavy bombardment!

5 IRON DISSOLVED IN THE OCEAN

The central role that iron plays in the functioning of life, I believe, has not been thoroughly examined by origin of life investigators. I am not just talking about the iron contained in the compounds that are the products of metabolism; these have distracted many investigators for years. I am talking about the iron that is incorporated in the structure of metabolism's enzymes; the enzymes that make metabolism possible in the first place.

Due to its transitional properties, this element can form more than four bonds. That is exactly what metalloenzymes require iron to do in order for them to form their enzymatic site complexes. Metalloenzymes are proteins that require metal ions to perform their enzymatic activities. The oxidative state that these iron atoms need to be in- in order to form these enzymatic site complexes- is exactly the same state as that of iron atoms dissolved in the oceans. Coincidence, I do not think so.

The iron ore that man excavates from the Earth comes from a large layer of iron ore that was laid down before large animal life started on earth. At least 3.7 billion years ago single celled lifeforms started using photosynthesis to make energy from carbon dioxide. This was a major milestone in the history of life, to be able to harness the light of the sun for biological energy. Here, however, the important thing to consider is the byproduct of that process; oxygen. Oxygen was released into the environment over billions of years. At the beginning every oxygen released was immediately reacted with every reactive element found in the ocean.

Eventually the iron located in the ocean increasingly reacted with the excess oxygen, precipitated out of the water, and was deposited in layers at the bottom of the oceans. The oceans contained so much iron that it took over half a billion years to form the banded iron ore formation; which can be found world wide today. To give you a better idea of the amount of iron diluted in the oceans back then lets look at the production of iron ore today. The annual production of iron ore world wide in 2015 was 3.3 Billion tonnes! This annual production, with no end in sight, has been going on for over a hundred years. Think about it: all the iron contained in every building, car, airplane, train, and ship around the world was once dissolved in the ocean. Imagine the amount of iron that had to be in the oceans to make that production possible and the amount of water in those oceans to hold it all.

An iron saturated ocean is not the conditions described by any origin of life research paper that I have ever read. None have addressed the reactivity of iron with any reactions they describe in their papers. Some of these papers only address the iron found in the byproducts of metabolism as they theorize how the environment could have created them. Then assume that life is going to reverse engineer the proteins to synthesize these products. That is not the approach I take in my theory. In Emergent Chemical Evolution, I not only account for the iron saturated ocean, but also describe how reactions are shielded from it, and how metalloenzymes take advantage of it.

All of the Iron dependent metalloenzymes require many dissolved iron atoms, at least eight, to form just one of their enzymatic site complexes. There is no other environment that can so readily supply so many iron atoms- and those of the right oxidative state- needed to take part in that complex. The fact that iron was readily available in the oceans is hard to ignore. That they are so central to so many proteins and metabolic pathways is also very hard to ignore. It should also be remembered, that in rock cavities and clay complexes the iron atoms would have been strongly chemically bound to the rocks and clay; making them unavailable for easy incorporation into enzymatic site complexes. This is yet another problem with cells forming in rock cavities and in clay complexes, something else that should be very hard to ignore! Iron's undeniable availability in the open oceans back then makes the open oceans the most likely site for the origin of life.

Section II
Difficulties of Other Hypothesis

6 LIST OF ERRONEOUS ASSUMPTIONS

The following list of assumptions, are those assumptions that origin of life investigators have been working with since the 1960's. I truly believe that these assumptions have, for all of the preceding decades, been limiting the progress of their research; hence my reason for calling them erroneous assumptions. Several of these assumptions fail simply from the logical fallacy of appealing to authority. Just because a Nobel laureate or other highly regarded researcher gives an assumption on a subject does not make it fact; Sure, due to their expertise and experience their assumptions should receive more weight and more consideration than others. But if over time, especially after 40 years, it has not been corroborated with experimental data or with observations then those ideas should, at the very least, be demoted. Here is a list of assumptions that, I know, should not only be demoted but outright discarded entirely and, in this section, I will give the reasons as to why.

The List of Erroneous Assumptions

1. Conditions of Pre-biotic Oceans
2. A Self-replicating Molecule was Initiator of Life
3. Lipid Cell Membrane was Necessary to Initiate Life
4. Virus Originated From Living Cells
5. Homeostasis Made Favorable Conditions
6. Life Occurred Randomly

7 THE CONDITIONS OF THE PREBIOTIC OCEANS

All previous articles that I have read regarding abiogenesis claimed that the prebiotic condition of the ocean's pH was basic. I have yet to see an article that challenges this assumption. Or that, at the very least, has shown that they have kept up with the latest advancements on the subject. Well, at least, all of the interviews and articles that I have read since I began researching this subject (2004). They all just state this claim as a matter of fact, and none have elaborated any further.

They did not offer any competing theories on the subject- probably because it was outside their area of specialization. Nor did they mention that there is no consensus among geologist regarding the conditions of the Earth during the Hadean era. Or, perhaps, the investigators just did not do enough research to know that there was no consensus on the matter. The previous section on Local Conditions of Prebiotic Earth, I believe, sufficiently elaborates on why this assumption has got to go; but, it would not hurt to quickly summarize them here.

The ocean is not a static body of water. It is and always has, since the beginning of its existence, been a very active body of water. Its ever present currents are always stirring up its contents, not allowing its content to stay in any one spot for very long. Also stripping any mineral cavities of their contents nor giving that content enough time, while in there, to react very much.

Back then the atmosphere contained much greater levels of carbon dioxide than it does today. The high level of carbon dioxide created pH gradients in the ocean making it easier for organic material to remain stable indefinitely. Modern life uses carbon dioxide to maintain its internal pH in order for its proteins not to denature or to dissolve; so, so much carbon dioxide was absorbed by the ocean- back then- that it effectively eliminated any dilution problems for life's first molecules.

The latest publications relating to the delivery of all of Earth's volatile inventory state that they were delivered by carbonaceous chondrites. Enough of these meteorites impacted Earth during the late heavy bombardment that they created our oceans. Among the content of these meteorites were over seventy different types of amino acids. The oceans were full of three times more types of amino acids than those that Life uses to make its proteins today. The extraordinary fact that, baring extremophiles, all of life uses amino acids as a starting point for making its most important molecules, to me, makes it abundantly clear that amino acids were all life had to work with at the beginning.

Iron's central role in the Krebs cycle is very hard to ignore also. The sheer number of Iron atoms needed for a single reaction in one step of the Krebs cycle calls into question the feasibility of any of the other origin of life hypotheses. Life is now capable of removing iron from rock, but that would have been a much harder task to perform prior to the emergence of cells. Iron's reactive nature with water makes the open ocean the easiest and most abundant source of iron for Life's origin.

8 A SELF-REPLICATING MOLECULE INITIATED LIFE

The assumption that a self-replicating molecule initiated life is an extremely pervasive one. It's the one that amazes me the most, because it has lasted so long. It is the key inhibitor of progress in the search for the solution to abiogenesis. It is so pervasive and so central to so many modern hypothesis of abiogenesis, that it has become the de facto starting point for all abiogenesis hypotheses. It was the main reason I waited to publish my hypothesis in a scientific journal.

If I would have tried to publish an article- that did not start with a self-replicating molecule- I fear it would have been summarily rejected. Many, I am afraid, cannot see beyond this misleading idea on the origin of life. And my unique take on the start of life, would have been exposed. Besides, it is abundantly clear that my hypothesis can only be properly presented in a book format.

It was also necessary to present my theory in a book format to be able to present all the information needed to comprehend it. It was necessary so I can properly combat the self-replicating-molecule assumption. It was also necessary so I can have enough time with the reader in order to undo the brainwashing this assumption has caused. And finally, it was necessary, to open the readers mind to the possibility of life starting in an altogether different way.

This assumption was proposed before we even had a thorough understanding of all the cellular machinery involved in the maintenance of DNA and cellular metabolism. Why would life abandon a simple system of singular molecular replication for one requiring many separate molecules? What evolutionary benefit would an organism gain by spending more energy maintaining extra molecules? Why do we not see any of these organisms today? Surely- if they were able to survive in that harsh environment back then- they would still be able to survive in Earth's increasingly stable and temperate environment.

Everywhere we look in cellular mechanisms, there are only molecules working in concert with other molecules. There are no molecules that replicate themselves, even partly. I have never come across any molecular system of replication that has any evidence of once being self-replicating. The solution to the chicken and egg problem is of course, according to evolution, the egg. But an egg it is, not from a chicken of course, but still an egg from another bird just the same.

They needed to look for a different bird to lay the egg of abiogenesis. They simply needed to look for a simpler system that replicates or creates biological molecules. They made the mistake of assuming that a completely different system was in place; a system that, until today, still has no evidence it has ever existed. Emergent Chemical Evolution not only proves abiogenesis began without self-replicating molecules; it does so without resorting to exotic non-life chemical reactions or extremely improbable environmental conditions.

9 LIPID CELL MEMBRANE NECESSARY TO INITIATE LIFE

The one thing, I think, that has escaped all origin of life researchers is how modern cells keep homeostasis. Sure you need a membrane to keep the outside out, but that is not its main purpose. Its main purpose is to keep the cytoplasm at a pH where its proteins would function, in our case of 7.2 to 7.4, obviously. That is basic biology 101, but here is the part that I think they have overlooked; they have overlooked exactly how it maintains that pH.

The membrane maintains the cell's pH by controlling the amount of carbon dioxide, in the form of carbonic acid, present in the cytoplasm. The acidic properties of dissolved carbon dioxide are what the cell uses to maintain this pH. The carbon dioxide content of the cytoplasm is controlled by an antiport protein that eliminates bicarbonate ions across the membrane. How was this accomplished by a membrane that was void of proteins?

Here, again, is another difficulty that modern hypothesis of abiogenesis do not account for. They posit that a membrane was necessary, in order to have the right conditions, for molecules to reproduce themselves. But they do not explain how the membrane makes those conditions possible? Or what mechanism would emerge to make those conditions possible within a membrane void of proteins? Hindered, by sticking to this assumption, they did not look for conditions that would

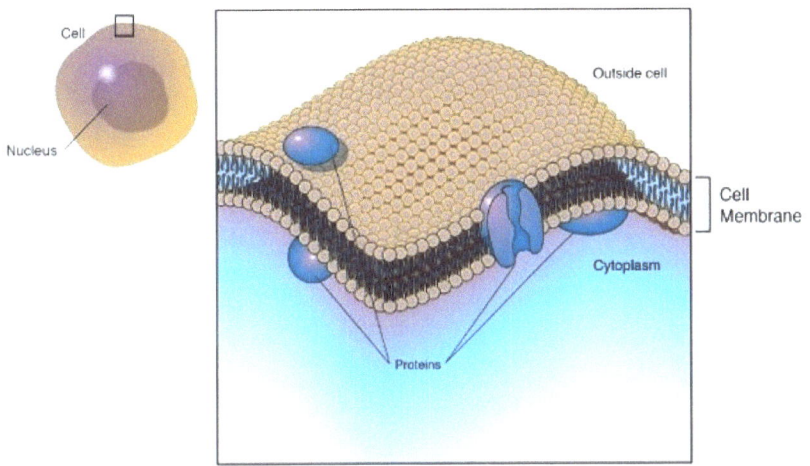

Figure 2. Proteins give the membrane its abilities.

allow polymerization to occur without a membrane. And, again, they could not look beyond this assumption. In essence, this assumption poses more problems than it solves.

The assumption is wrong for several reasons. The first reason is that with no mechanism to bring in fresh material inside a double emulsion bubble to replicate the molecules, evolution could not take place. The second reason is that without a mechanism in place to hold and repair the membrane it would eventually pop like a balloon. Spontaneous double emulsion bubbles would not last very long. Its contents would then be easily released back into the environment before any meaningful reactions, or evolution, could take place.

The third reason is that what would keep the double emulsion bubbles from merging into a larger bubble? The bubbles would not have any structural proteins to maintain the integrity of the membrane or to maintain the integrity of the cell, itself, for that matter. So, void of any means of keeping the cell together, the protocells would merge into larger ones, and would have eventually exploded. The fourth reason is that after years of making double emulsion bubbles in perfect laboratory conditions, they have not been able to make them last very long; it seems highly unlikely that they would have lasted, at all, in the physically tumultuous prebiotic conditions.

The fifth and final reason is that conditions on the earth have always been very stable. The earth has maintained its composition and thanks to its gravity its atmosphere. So conditions on the Earth have never changed very drastically. There was no need to protect the molecules from the environment back then because of this fact. At a molecular level there was no need to keep homeostasis due to the chemical stability of the environment.

Remember this- the last point- because it plays a very key role in my theory of Emergent Chemical Evolution. As we go through the list of erroneous assumptions the problems with the assumptions will point you to the solution. The point of first going through the list and then breaking down every item is to prepare you to see how ECE answers these problems; as well as to show why these assumptions are not only obsolete, but wrong.

10 VIRUSES ORIGINATED FROM LIVING CELLS

This one assumption has not only limited the search for the origin of life but also the science of virology itself. The name- virus itself- is derived from the effect they have on living cells. This form of "life" was discovered because of its power to cause disease. The way they take over cells and destroy them is another reason why some humans despise them.

The fact that viruses need to reproduce inside a living cell and sometimes destroy it, in the process, is what really offends our sensibilities. There is something, about this fact, that scares most people, and, I guess, most scientists. This fact alone makes it hard for scientists to think of this life form as anything more than just a parasite.

It has long been argued whether viruses are alive or not. The fact that they require a living host in order to replicate is one of the main reasons it has been argued that they are not. That they do not respire or mobilize are also very powerful reasons not to believe they are alive. All of these reasons also place viruses in between what is "living" and what is not.

Viruses nicely fit in that void that exist between fully inanimate matter and living cells. They are very resilient pockets of organized matter. They are nearly indestructible, able to survive in all types of hazardous environments. And once they find suitable conditions they become active and begin replicating uncontrollably. That seems like a very useful trick for a non-living thing to have. The question is how did that useful trick evolve? A better question yet is how did viruses evolve in the first place? An even better question still is what mechanism really gave rise to viruses?

What mechanism has ever been observed that produces an original virus? Has anyone ever induced a living cell to produce an original virus? What has been observed when genetic mechanisms go out of control? Cancerous cells are produced or the cell dies. Also there are very different types of viruses that use different types of RNA and DNA systems altogether. If viruses originated from living cells, all of which use the same system, how did these other viruses evolve?

There is one major fact about viruses that should have been a red flag for researchers to know that cells did not give rise to viruses. This fact is that viruses can self assemble in a petri-dish. Place the individual molecules of a virus into a solution and without the aid of any other molecules they self-assemble into the complete virus!

Cells on the other hand do not self-assemble in any circumstance. Once the cell membrane is breached rarely does the cell survive. So taking its individual molecular components and placing them in a petri-dish has never resulted in a cell spontaneously self-assembling. This clearly places the origin of viruses before that of cells. If a protocell could form in the harsh conditions of the prebiotic Earth, why would not a modern cell in perfect conditions be able to reassemble itself?

Well, Emergent Chemical Evolution can answer all of these and many more unanswered questions about what we see in viruses and in life today.

11 HOMEOSTASIS WAS NEEDED TO START LIFE

The assumption that life needed to invent homeostasis in order for it to thrive is untenable. How did the conditions for evolution emerge in the first place in order for homeostasis to evolve? There is a great divide between the membrane and the genetic information of the protocell. How would the appearance of a membrane cause its genetic information, for its assembly, to spontaneously appear? Let alone the large amount of genetic information needed to create a system of homeostasis. That these extremely complex things would have happened spontaneously goes against everything I understand of evolutionary principles.

In this scenario, there would be no evolutionary pressure to add additional genetic information in order to produce a membrane, and as well a system of homeostasis. Expending extra energy to maintain additional genetic material and having a redundant system would put the protocell at a disadvantage. Whereas its competition would not have to expend any additional energy; and therefore, would require less energy from the same environment. This would put any cell- with this additional genetic baggage- at a disadvantage and would quickly become extinct according to evolutionary principles.

It also was not necessary to figure out how an area

with conditions like those found inside a modern cell arose; in order for a self-replicating molecule to form, then create an unnecessary membrane and an unnecessary system of homeostasis. There are modern single celled life forms that live in very harsh environments, in deed, ones that would kill alot of other modern organisms. Why must we assume that those extremophiles represent the first life forms that emerged? Why assume that they emerged in those environments and not that they evolved to occupy them? What makes more, evolutionary, sense is that modern cells evolved homeostasis to adapt to an ever increasingly hospitable environment, for us, but an increasingly deadly one for them, the first life forms.

Evolutionarily speaking, it would have been much simpler for a membrane to evolve after a system of replication had evolved first in a conducive environment, but not one conducive to modern life forms. In a conducive environment, the membrane would first evolve to serve as a means of protecting the individual's resources from competing life forms; hence no need for it to maintain homeostasis. Then slowly it would begin to evolve a system of homeostasis as the environment slowly changed into a more hostile one. And we can be perfectly certain that it did change- breathing in oxygen while we read this- and that it did so slowly, by observing the banded iron rock formations. Which are evidence of the slow oxidation of the atmosphere over two billion years.

With debunking this assumption, we finish explaining why a membrane was unnecessary at the beginning of life. In order for homeostasis to work, an organism needs a boundary between itself and the outside world. This boundary is only necessary, for homeostasis, if the conditions of the environment surrounding said organism are sufficiently different from what it needs to reproduce itself. Again, Emergent Chemical Evolution will conclusively explain how a membrane, first, was not necessary but later would evolve, and then show how a system of homeostasis would emerge after the environment started changing.

12 LIFE OCCURRED RANDOMLY

The assumption that Life occurred randomly is a forgone conclusion if you subscribe to all of the previous erroneous assumptions. These assumptions lead to a very different view of how life should have arisen: first reason for this erroneous conclusion, is that the conditions in the oceans did not favor life; second reason, having no idea what conditions in this environment would have favored the emergence of a self-replicating molecule; third reason, having no idea how to get said self-replicating-molecule and its supplies inside a protein free membrane; and the fourth final reason, having no idea what mechanism a protein free membrane would use to maintain homeostasis; hence you have the erroneous conclusion, that the emergence of life was a random fluke of nature and that we're very lucky to be here. Wrong!

Please do not get me wrong, I do believe we are lucky to be here, and that we are here by chance. But, instead, what I am saying here is that I disagree with the notion that life started here with one random improbable event. Instead, I posit that Life emerged here from the constant physical forces that have always existed in our Universe, at the very least since the formation of our solar system.

With my explanation of local conditions in the prebiotic oceans and the debunking of all these erroneous assumptions, I hope a clearer picture of the path to life is forming in your mind's eye. If not, do not worry the next section will certainly do the trick.

Section III

Emergent Chemical Evolution

13 PUTTING IT ALL TOGETHER

Today it is well known that all chemical reactions performed by Life are done using enzymes in water. It is also well known that amino acids, which make up all enzymes, spontaneously organize into alpha-helix structures in water. But apart from what has been well known I found a peer-reviewed published article that shows amino acids spontaneously bond to each other in water. So when we realize that the Earth was covered in water before Life started makes the theory Emergent Chemical Evolution a better explanation for the origin of Life.

In the first section of this book, I clearly established that the conditions on the prebiotic earth were not like how many had previously imagined. In the second section, I also clearly established how all the other hypotheses of abiogenesis are flawed; I did this by pinpointing all of the erroneous assumptions that they used as the basis for their research. They are not just erroneous factually but logically as well. Those assumptions all suffer from the same logical fallacy; the fallacy of appealing to authority. Today's researchers relied too heavily on the conjectures made by noble laureates four decades earlier. In using the hypothesis of those noble laureates, they skipped the first step of the scientific method; which is where you observe nature. When I embarked on this mission I decided to revisit everything; I was not going to have any "sacred cows" pooping all over my efforts.

What remains, for me to do now, is to correct all of those errors and then answer the one fundamental question. The question that everyone would like answered- exactly how did Life begin? If all those other hypotheses are incorrect, then what is the right one? Which one can also explain how life became the way it is today?

Having the power to explain how life became the way it is today has got to be a key component of any correct hypothesis. A correct hypothesis must have the power to not only explain what we see, but also be able to predict what we do not. A correct hypothesis should logically follow from any findings- past or present. It should also leave, in the reader's mind, a clear and vivid picture of the process from beginning to end.

My Theory - Emergent Chemical Evolution- I know can answer all of these questions and meet all of those criteria. Now the only thing left for me to do is to explain Emergent Chemical Evolution in a logical and cogent manner, but in order for me to do that you must have read and understood all of the previous sections of this book. And if you did not understand the local conditions that led to the mechanism of Emergent Chemical Evolution (ECE) or the difficulties of the other hypothesis, I take full responsibility for not doing a better job. But at this point you have only two options. One is to stop here and re-read them until you do. Or two continue reading and then, when you are done, return to those sections with the foreknowledge of their future relevance which might prove even more helpful in understanding them.

Either way I sincerely hope that- by the end of this book- you understand my theory completely; see its logical flow; examined its modern published evidences; and then once you have realized its simplicity that you can see its inevitability of producing Life here on Earth.

14 STRUCTURE NOT ENERGY IS MOST IMPORTANT

Structure is the key defining trait of life. Structure is because life uses normal atoms; same as, the rest of the universe. Structure is because- contrary to popular belief- the processes of life do not violate any laws of nature; same as, the rest of the universe. Structure is because- even though they use the same atoms- what separates inanimate matter from Life is Life's molecular structure.

    What has hindered origin of life researchers for all this time is picking the wrong level of Life's organization- ie structure- to study. Some aimed too low trying to figure out how RNA formed from simple compounds. Others aimed too high trying to make protocells from simple compounds. The problem with both of these strategies is their focus on simple compounds on the assumption that that was all there was on earth back then. What they needed to do was look for slightly more complex compounds. Compounds that are: spontaneously formed in nature; are ubiquitous; are modular; easily link up to each other; when united spontaneously form structures; and most importantly can use themselves as a starting source to build different and more complex structures. What I have just described and which are at the proper level of Life's organization are Amino Acids!

    Structure is, because proteins can do the things they

do from the shapes they take and not just from the chemical properties of their individual amino acids. Amino acids mostly help to form the final shape of the protein; for example, amino acids that are hydrophobic become part of the inner structure of the proteins. But the one feature that unites all amino acids is a common arrangement of atoms. This arrangement, or structure, allows them to be joined into chains; again, structure not chemical properties is more important.

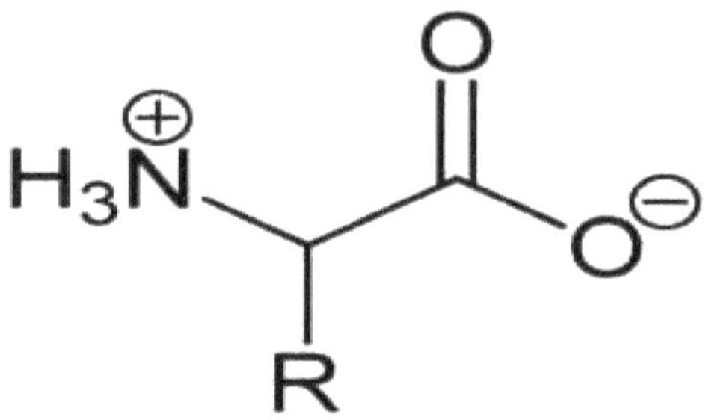

Figure 3. The Common structure of all amino acids that allows them to easily bond to each other and form long chains.

    The structure of the protein is what DNA and RNA code for. DNA nor RNA give instructions- as in a computer program- on how the proteins should behave or tell them what products to produce. DNA and RNA are structures that function as templates- which are structures too- that determine the amino acid sequence of proteins; therefore making proteins of a particular structure. A structure that causes them to make a certain product or behave a certain way. Proteins, with their structure, allow life to do what life does. Proteins, with their structure, are what give life its shape. Proteins, with their structure, facilitate chemical reactions for life. Proteins, with their structure, give membranes their function.

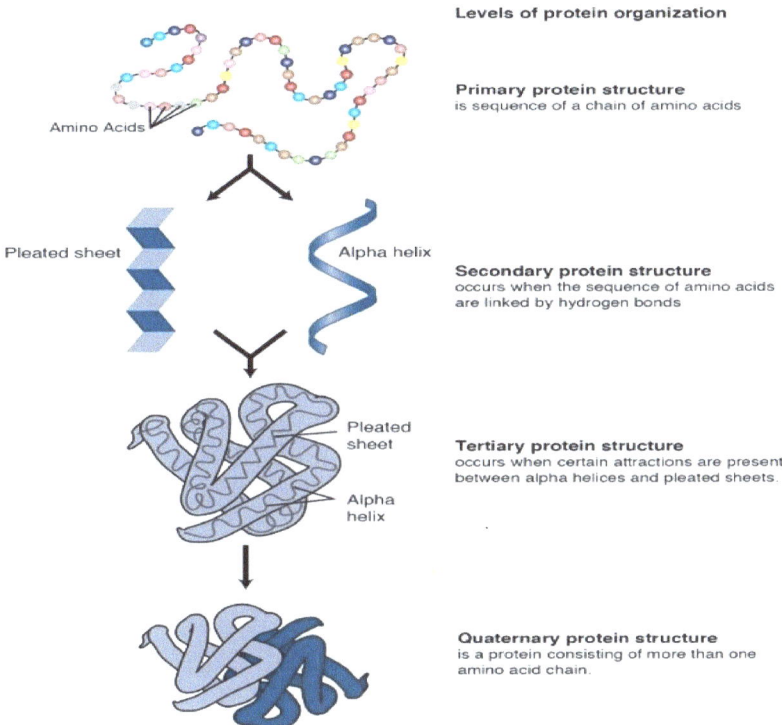

Figure 4. The structure of the protein is what finally determines if it will function as an enzyme, scaffolding, or a cofactor.

The protein that copies RNA is different in the different branches of life. Bacteria, fungus, and humans all have proteins that copy RNA. These proteins perform the same function; copy RNA and make messenger-RNA so that proteins can be made from that RNA template. Each one of these different life forms has a version of this protein. The amazing thing about these different proteins, that perform the same function, is that their DNA that codes for them are vastly different; up to 75% different. The proteins that are produced are composed of vastly different sequences of amino acids. In some cases up 75% of the amino acids are different. What is the same and allows them to perform the exact same function is their shape; their structure. The amino acids ability to form repeating structures allows these vastly different sequences to form the same structure and form the same reactive regions. That is why structure is more important than energy, because it is their structure that allows these different proteins to perform the same task.

Membranes and their proteins keep the proteins inside the cell separated from the rest of the universe. Proteins are able to do these things and much, much more thanks to their shape; their structure. The flow of energy in living systems is more complex than in inanimate matter; but, this is due solely to life's more complex structure. The flow of energy does not drive the chemical reactions necessary for life. It is the shape and structure of proteins that drive the flow of energy in, out, and within the cells of living systems. The flow of energy and chemicals, within the krebb cycle, are just byproducts of the proteins within the mitochondria and chloroplasts. Trying to figure out how life started by just analyzing the flow of energy and chemical byproducts within the Krebs cycle without analyzing its proteins; is like trying to tell the year, make, and model of a car by just analyzing its exhaust. It just cannot be done.

Even if you knew the amount of fuel that is not combusted by a particular car. You would have to know what brand of fuel the driver put in the car. You would have to know if he ever used a different brand of fuel, even just once and how much of it. You would have to know if the car is perfectly tuned or not, and if not how out of tune. You would also need to know: how long the car was running before the sample was taken, what was the ambient humidity and temperature; And, after figuring all of that out, you still could not know- with any real certainty- what car made that exhaust.

By explaining Emergent Chemical Evolution I describe a mechanism that would lead to the emergence of these structures. And it is the interaction of these structures that would eventually lead to the emergence of life, with no protocells in sight.

15 THE WORLD AS A SYSTEM

Although I am confident that I have thoroughly debunked the whole idea of protocells being the ancestor of modern cells, I believe that the facts, concerning the conditions of the prebiotic Earth, lend themselves to a very useful metaphor: and it is, that the entire Earth functioned as a protocell. The protocell Earth provided the environment and conditions necessary to bring forth the mechanism of Emergent Chemical Evolution (ECE).

It was the atmosphere of the Earth, causing pH gradients in the oceans, that made the oceans the functional equivalent of cytoplasm making the entire Earth a protocell. The entire planet Earth would have functioned as the first and only protocell. A couple of weeks before publication, I came across evidence in support of my conjecture; that the pH gradient of the oceans would lend itself to this metaphor. What this article states is that modern cells set up pH gradients in their cytoplasm too. According to Jaroslov Flegr modern cells use this pH gradient to increase the effectiveness of their enzymes (Flegr 2009). So having a pH gradient is not only very good for modern cells, it can also be reasonably surmised that it would have been very good for ECE as well.

The Earth, as a whole, would have been keeping things in balance- as is its tendency to do so today. Much the same way a cell keeps things in balance within its membrane. So does the Earth; by inducting material into the mantle, by releasing material through volcanoes and erosion, and by maintaining all manner of other cycles, for example: the carbon cycle, the water cycle, the convection cycles, etc. With all of these physical cycles working together to move material in and out of the oceans, the oceans became the functional equivalent of the cytoplasm of the Protocell Earth. This cytoplasm- literally the oceans as a variably acidic solution- would have made the entire surface of the Earth, which was covered in water- all of it being equally pressed upon by the carbon dioxide laden atmosphere- was favorable for ECE.

Thinking of the Earth in this manner was only possible after I had gained enough knowledge of chemistry, physics, astronomy, geology, meteorology, oceanography, and biology. Only then was I able to see how the prebiotic Earth's systems would have interacted. How they would have created the necessary conditions for the emergence of ECE. And, I am sure, it was only my intense interests- in such a wide variety of scientific fields of study- that led me into having this multidisciplinary view of the prebiotic Earth.

Unfortunately, by necessity, all fields of scientific study are becoming more and more specialized. This is only making it harder for scientists to get a broader perspective on problems that are larger in scope than that of their individual fields. All the skills they need to get ahead in their specific field of study are not enough to solve this big of a problem. I believe my love of science, and my passionate interest in such a wide variety of subjects, has served me well in gaining the needed broader perspective to solve this big of a problem.

Another advantage that I have, over other investigators, is my intense interest in and study of chemistry since I was a teenager. It has helped me solve many complex problems during my twenty five years of experience as an electronics technician. I was able to go beyond just theoretical physics and what a regular technician, one without a chemistry background, could have done in understanding how the electronic components worked. All of these years of "working" experience, using what I know about physics and more importantly chemistry, in solving not just complex problems, but messy "real-world" problems, gave me- I believe- a real advantage over the other origin of life researchers in solving this problem.

So when the company I worked for as an electronics technician was shipped to Mexico, I took the layoff and returned to school to become a nurse. While I was taking the anatomy class to become a nurse, I learned something that would lead me to my second breakthrough (another Aha! moment); and eventually to ECE. I learned that if our blood's pH level goes out of the range of 7.2 to 7.4 we can become very ill. But the most important and surprising thing I learned, that day, was that not having any carbon dioxide- at all- in your blood is very, very bad. The very important reason for this is that our bodies use carbon dioxide to help acidify our blood and cells; thus, keeping them within that healthy pH range.

When I learned about carbon dioxide's role in keeping our homeostasis, I immediately thought about carbon dioxide's role at the beginning of life. I asked myself, how could a protocell have adjusted its pH? I also knew that there was no free oxygen in the atmosphere back then. So how did all that carbon dioxide in the atmosphere effect the environment and the first molecules that lead to life? So in order to answer these questions, I began to study carbon dioxide and its effects on living cells, water, and the environment. And what I learned, led me to the conclusion that protocells could not have existed; being void of proteins to regulate carbon dioxide and thus its internal pH. That it was not protocells that gave rise to life but, instead, it led me to a better conclusion: that the Earth itself acted as the first and only protocell. There are many reasons why I do not just say that only the oceans acted as the protocell. These reasons are that you need the Earth's gravity, atmosphere, spinning, and mantle to achieve the proper conditions in the oceans in order for those pH gradients to have formed and thus ECE to emerge. The oceans alone could not have initiated Life.

16 THE CURRENTS OF LIFE

Emergent Chemical Evolution (ECE) was initiated first by the currents activating the amino acids. The currents achieved this by passing them through the ocean's pH gradient because different amino acids become charged at different pH levels. And because the amino acids were activated in this way they readily bonded to each other forming polypeptides. Forming polypeptides while still riding the oceanic currents. Now these, polypeptides, would be composed of random sequences of amino acids. And this would have given them the ability to catalyze innumerable types of chemical reactions; therefore, enzymes were available first to catalyze all the chemical reactions needed to initiate life.
    What? What just happened? Hold up! Wait a minute! Now that is one really big jump from amino acids to proteins. Now how did that happen? What mechanism was at work that caused amino acids to spontaneously form into proteins?
    Well I can explain: when I remembered that meteorites carried amino acids from outer-space and that the small amount of water they contained created the oceans. I then came up with the solution that the amino acids would have been present in the oceans, too. And I also thought that they would have spontaneously polymerized and become the first proteins; hence, they would be the first enzymes that would have catalyzed all

the reactions necessary to initiate Life. Eureka!!!

With the unending currents turning up the contents of the oceans, that content inevitably passes through all of the depths of the oceans- with each layer having its own pH level. The first thing I have to establish is the identity of this content. Which- based on the latest published studies (Martins et all2007, Alexander et all 2012)- I concluded were amino acids, dissolved iron, meteoric phosphorous, and other trace metal ions. These are obviously not the only constituents of the oceans at that time but for the purposes of this hypothesis I consider them the most essential. They play the most fundamental roles in the initiation of life. They are the starting constituents of ECE.

The first constituents we need to look at are the amino acids. The amino acids are the most important because they are organic molecules. These are the same organic molecules that life uses to build its structure and to perform all of life's chemical reactions. These organic molecules are also subject to the rules of organic chemistry. And in organic chemistry the organic reactions and bonds that are formed between molecules mostly occur at the reactive regions of these molecules and ions.

These reactive regions on the molecules generally fall into two types. One type is the negatively charged high electron density site of a molecule that has unshared electrons. These electron rich regions are called nucleophilic and the molecules possessing such regions are called nucleophiles or electron-donors. The second type are the positively charged electron-deficient region of a molecule that are known as electrophilic and the molecules that have them are called electrophiles or electron-acceptors. Most organic covalent bonds are formed by the coordinated uniting of a nucleophilic region of one molecule with the electrophilic region of a second molecule. Once these two regions come near each other they spontaneously attract and unite forming a covalently bonded larger molecule.

Now here is the take home message of this small organic chemistry lesson; every amino acid has both nucleophilic and electrophilic regions on them. They are known as zwitterions- molecules that have both a high electron density region on one side and an electron-deficient region on the other; making them modular. All amino acids have either one of these regions active when dissolved in a solution that has a pH within- the very wide- range of 2 to 9 pH. This means that amino acids would naturally and easily form covalent bonds with each other whenever they came near each other at the correct depth(pH level) of the ocean. No exotic chemistry or conditions necessary for the origin of proteins; or Life for that matter. As a matter of fact all the proteins, from all branches of life, that make RNA start by transforming an amino acid to start the process. Now that I think about it all important molecules in life have their start as a transformed amino acid.

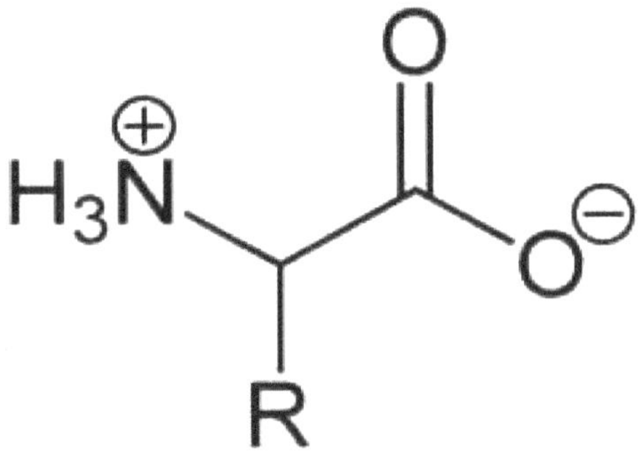

Figure 5. **The amino acid can make a bond with another amino acid on either side of itself throughout the very acidic 2 to the very basic 9- pH range.**

I immediately thought, how am I going to prove this? Is there any way that I could quickly and cheaply test these ideas? And in my home? Or was there anyone that I could contact who could get me access to a professional laboratory? Then I remembered that, in science, one should first do research and see if anyone else has answered this very question before. Or check and see if anyone has done any experiments with amino acids in water. I decided to do the necessary research online first and avoid reinventing the wheel.

The next time I was sitting in front of a computer, I got on the internet and did a Google search for the words "Aqueous Amino Acid Polymerization." I was very proud of my choice of words for the search. I knew that this choice of words would eliminate most of the search result junk. And when I hit enter- I hit pay dirt. I found an article published in The Journal of Biological Chemistry way back in 1953. As it turns out R. R. Becker and Mark A. Stahmann discovered that amino acids in a buffered solution would prefer to spontaneously bond to each other, rather than to the water molecules that were present in 5000-fold excess. Wow! And according to the paper, they also spontaneously bonded to each other most efficiently at or near the neutral pH of 7.2 to 7.4. Boom! Which is the exact same pH as that of the cytoplasm contained within living cells (Becker and Stahmann 1953). EUREKA!!! I found it and I was right!!!

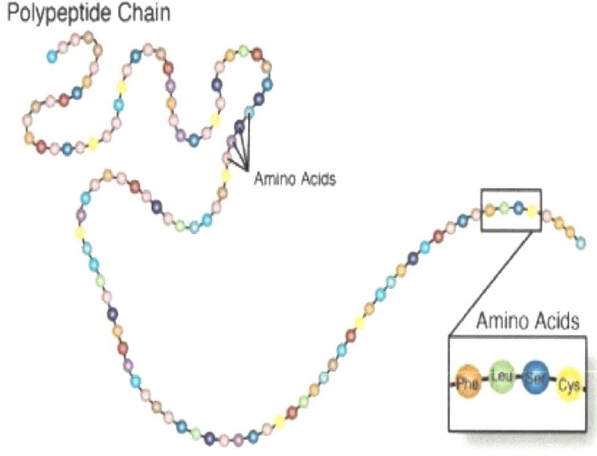

Figure 6. Amino acids form polypeptides spontaneously in a buffered water solution even when the water was at a 5000-fold excess.

I had just found the evidence that I needed to prove that amino acids do spontaneously join together in water. Knowing that the amount of carbon dioxide in the atmosphere back then would have buffered the oceans, it was almost more than I could bear. It was all I could do to keep from jumping up and down yelling "EUREKA!!!" in the middle of Miami Dade College at the top of my lungs.

Now- I thought at that moment- there is a big difference between the method used by the researchers to control the level of pH in their experiments and the method used by living cells. In their experiments the researchers used large molecular weight chemical buffers to control the pH in a small container. The carbon dioxide released by the chemical reactions inside such a small container would quickly overcome the buffering of the solution; thereby not allowing longer polypeptides to form (Becker and Stahmann 1953). I posit that the prebiotic oceans would have acted as the largest buffered solution imaginable and would not have been so susceptible to its pH being changed so easily; thus allowing larger polypeptides to form in the prebiotic oceans, as they do so in cells today; which are solely buffered by carbon dioxide.

Whereas a living cell controls its internal pH by controlling the amount of carbon dioxide, in the form of carbonic acid, contained within its cytoplasm. My conjecture is that carbon dioxide is nature's smallest and most readily abundant molecular buffer. Which its small size makes it easier for larger polypeptides to form within living cells. And large polypeptides, as I theorize in this book, would have also formed in an ocean environment buffered by the atmosphere; which was then laden with carbon dioxide, nature's smallest and most readily abundant molecular buffer.

The carbon dioxide laden atmosphere would have made the oceans a variably acidic solution; making the entire surface of the Earth, which was then covered in water, favorable for ECE. Think of it, millions of square miles of variably acidic solution with trillions upon trillions of amino acids linking; forming longer and longer polypeptides(proteins/enzymes). Millions of square miles of all the known elements, molecules, and their ions; always available and incessantly brought back by the currents; to react again and again with the ever growing proteins. And as the polypeptides pass through the layers of pH they would have also become denatured. Untangling the polypeptides thus making it easier for them to gain additional amino acids. But this was not the only means by which polypeptides could have formed and grown longer in the oceans of the prebiotic Earth.

Have you ever wondered why water that collects at construction sites turn green or turquoise? The answer is chemically complicated but easily explained using everyday terms. The water, we see pooling by the roadside, basically rusts the iron out of the rocks and soil that are holding the water. Water is nature's most versatile and potent solvent due to water's chemical structure. Water can easily rust, dissolve, or erode almost every metal or compound found in nature. Knowing this property of water and that the Earth was covered in water before life started; what else could we reasonably expect polypeptides suspended in the water to do?

We could only reasonably expect polypeptides to have bonded to dissolved iron and other metal ions that saturated the oceans back then. I cannot see the polypeptides not reacting with these ions when they both were dissolved in the same water. I conjecture that this was another means by which the polypeptides could have lengthened themselves into proteins. It works like this, metal ions which are transitional atoms- and this makes it easier for them to do this- would bond with more than one amino acid or polypeptide. Then the metal-ion-amino-acid complex would be carried off by the currents and then when it passes through a certain layer of ocean, that has the necessary pH, the metal ion is removed by an electrophilic attack on it. And then once the metal ion was gone the amino acids would have bonded with each other; thereby making a longer polypeptide.

Today, we all know, iron is necessary for many metabolic enzymatic reactions, and that without iron we cannot transport oxygen from the lungs to the individual cells. We also know that Iron does these things by using its transitional metal properties to form metabolic enzymatic complexes; or in the case of hemoglobin by uniting four different proteins to oxygen atoms. And here- I present to you- the observed modern evidence of iron's ability to unite polypeptides in an aqueous solution. And this, I also consider, is additional modern evidence in support of the mechanism of metal ion bonding. Which also plays a central role in my theory.

The current theories on the origin of the oceans all agree that the oceans arrived due to the late heavy bombardment. That the meteorites, asteroids, and comets delivered the water to earth. Now the delivery of the water is a very important event for the origin of life but it is not the most important. Recent experiments done by Jennifer G. Blank, Ph.D., and colleagues at the Bay Area Environmental Research Institute NASA/Ames Research Center, Moffett Field, California showed that the impacts of these bodies also combined the amino acids found in them into peptides. Now, her experiments were only meant to show if the amino acids contained within the comets would survive the impact against the Earth. But she placed amino acids in water and impacted the water solution with a bullet travelling at 5,000 miles per hour (Blank). This experiment not only shows what happens to amino acids contained within comets but- I here posit- the experiment also showed that amino acids, floating in the oceans, would have been combined too when they were impacted by a meteorite that might not have been carrying any amino acids.

So not only did these impacts not destroy all of the amino acids, their energy also combined them, too! Solely judging from the amount of water contained in the oceans, an uncountable number of objects did slam into the earth; and every time they did so they united amino acids into peptides and those peptides into polypeptides. Again, judging by the amount of water in the oceans, it must have been a hell of a lot of impacts; therefore a hell of a lot of polypeptides, too, dare I say proteins by the end of it.

The shear number of objects hitting the Earth is not the only thing that needs to be considered here. These objects ranged in size from a few feet to hundreds of miles. The surface area alone of each of these objects would have bound trillions, at least millions, of peptides upon impact. Even if the meteorite was just a few feet. It would have traveled at such high velocities that the shockwave, when it struck the water, would have traveled hundreds of miles; potentially binding trillions of peptides along the way.

I would also like to draw your attention back to all the amino acids, polypeptides, and iron ions that were in the oceans back then. And I would also ask you not to forget that the currents of the oceans were and always have been relentless. The currents kept taking those growing polypeptides round and round through endless cycles- recursive cycles if you will. Now, this is a key step in the process of ECE because those relentless currents set up Emergent Recursive Cycles (ERCs) in the prebiotic oceans.

These ERCs would have continuously carried those polypeptides with them. Every cycle these ERCs completed added more and more amino acids and other polypeptides to the innumerable already growing polypeptides they carried. Those polypeptides would have kept growing longer and longer. Then, depending on the order of the polypeptide's amino acids sequence, the conformation of the final product- the protein- would have been determined, and would have then acted as either a structural protein, a co-factor, or even an enzyme.

17 THE CHAOS OF ECE

Emergent Recurrent Cycles (ERCs) remind me of those simple reiterative programs described in that great book *Chaos* by James Gleick. I read in that book how scientists and mathematicians used reiterative programs to model nonlinear systems in nature. I saw marvelous pictures in this book of graphic representations showing the results of these simple repeating programs. The complex patterns in these pictures mimicked the structures that are observed in nature: ferns, jellyfish, and the cloud patterns of Jupiter (Gleick 1987). And just like the reiterative programs described in that great book I see ERCs as being nature's ultimate simple reiterative program; because the ERCs' end result was the ultimate output of all, the emergence of Life itself.

The behavior of ERCs reminded me of this one paragraph that James Gleick wrote in his book, *Chaos*, that always stuck with me and, I believe, best sums up why ERCs would have been so effective. For me to try and paraphrase such a profound message would only serve to destroy its elegance and eloquence. So here is the direct quote of that paragraph:

"In the intervening twenty years, physicists, mathematicians, biologists, and astronomers have created an alternative set of ideas. Simple systems give rise to complex behavior. And complex systems give rise to simple behaviors. And most important,

the laws of complexity hold universally, caring not at all for the details of a system's constituent atoms."

In ECE the constituent atoms were the amino acids and metal ions dissolved in the oceans. The simple system was the ERCs cycling through the pH gradients of the oceans. The complex behavior was the production of proteins, enzymes, and RNA segments. This gave rise to the complex system of RNA-to-Protein translation which in turn produced its simple behavior of producing self-assembling viruses. And these simple viruses gave rise to the complex behavior of evolving evermore complex structures.

I have always been attracted to chaos theory for the beautiful images that were made by the computer programs written to demonstrate it. I am still particularly impressed with the images they produced that mimicked nature. Impressed by how one set of instructions repeated over and over again could reproduce the complexity of natural forms. Back then I felt that chaos theory would be the key to finding the theorized mathematical equation of everything. Now I feel that it could be the key to demonstrating mathematically how my theory would produce useful proteins by simulating what sequence of amino acids would be produced by ECE.

Before we go to the next chapter, let us look for an example of chaos theory in nature. We will look at, what some in colder climates might consider mundane; the formation of snowflakes. Before we even learned the physics of snowflake formation, we learned that no two snowflakes ever look alike. Why is that? Because of the snowflakes constituent atoms- in this case water molecules. Water molecules are, just like amino acids in the prebiotic oceans, ubiquitous in the ocean of air that surrounds the Earth. These water molecules are taken by the wind, the currents of air, up and down the temperature gradients; just like amino acids in the prebiotic oceans. Water molecules, just like amino acids, have a positive side and a negative side. So when a water molecule attaches to another water molecule, just like amino acids, it connects its positive side with the negative side of the other molecule. This simple repeating mechanism produces the most beautiful, complex, and unique patterns of snowflakes. In the formation of snowflakes we do not only have a natural example of Chaos theory at work, but also a perfect scientific equivalence for the ERC mechanism of Emergent Chemical Evolution.

18 THE INEVITABILITY OF LIFE

Emergent Chemical Evolution would have undoubtedly produced an innumerable amount of molecules, but how many would have been biologically important? What is the probability of a sequence of amino acids forming a biologically active protein? One, or many, that would be useful for the origin of life? In the first draft of this book I wanted to avoid using complex mathematics and long drawn-out explanations of complex concepts. In this final draft, I not only learn that that was naive of me, and impossible, but that they would prove my theory, and- given the conditions for ECE- would show that life was inevitable.

   Trying to prove anything mathematically is hard enough in its own right. So imagine how difficult it is to prove, mathematically, something that has never been witnessed nor has much evidence. Imagine trying to prove something mathematically when you are not even a professional mathematician; when you are not even sure that any math exists that can be used to solve this problem; and on top of all that- which makes it even worse- trying to prove a concept you came up with; that, as far as you know, no one else has ever proposed, let alone proven mathematically; So, it seams, complex and long drawn-out explanations are the order of the day.

   I have: four books on human anatomy physiology, three on physical chemistry, three on organic chemistry, two on biology, one on biochemistry, and one on molecular cell biology; all university level textbooks, and all of which point out that the number of different combinations of amino acids possible for a small protein of seventy amino acids ($20^{70}= 1.18 \times 10^{91}$) equals more than the calculated number of atoms in the universe (4×10^{81}). This is a fact that is never going to change, but it is not an insurmountable fact. All these textbooks address: the folding and shapes that these strings of amino acids (proteins) take; mention that they form active sites; and that they- for the most part- are catalyst. What they don't- because no one has until myself- address: is the relationship of the location in 3D space of each amino acid residue, that form the active site, with respect to their sequential place in the sequence of amino acids that form the protein. This relationship not only proves my theory right but also shows that- on Earth- Life was inevitable.

Before we can get into the details of this relationship, we must first understand how they arrived at the number of different sequences that amino acid residues can take in order to form a single length of protein. This will serve: both as a basis for understanding why they thought it was an insurmountable challenge, and also as a basis for understanding the 3D-amino-acid-residue relationship.

The amount of possible sequences that can be made from a chain of amino acids is calculated by using powers of twenty. Twenty because life only uses twenty amino acids to make all of its proteins. The power used depends on the length of the polypeptide (protein) being calculated- every amino acid adds one power to the calculation. For example, say you have a polypeptide of one hundred amino acid residues the number of possible sequences is calculated by taking the number of possible amino acid residues, twenty, raising it to the power of how many residues are on the chain, in this case one hundred; which gives us: $20^{100}=1.267\times10^{130}$ which is clearly more than the calculated number of atoms in the universe.

I can only assume that their reasons for pointing this out was to show the specialness of Life; to show the probability of a specific enzyme forming at random in nature; to show how difficult it is to decipher how proteins form enzymes; or maybe, to show how every protein is unique and how difficult it is to design a protein sequence with a specific enzymatic property.

It might have been hard for humans to come up with different sequences to form specific enzymatic properties but it was not so for nature. In nature, across Life's different kingdoms, for every enzyme found in any one kingdom there are more than twenty different sequences of proteins, across all of them, that make an enzyme with the same exact function (a homolog); and curiously enough with very similar shapes. So in reality, the odds are not as bleak as one might think. In contrast, according to a concept that I came up with concerning a protein's active site's orientation and its amino acid sequence, the odds were very good for the emergence of biologically active enzymes via Emergent Chemical Evolution. But before I can explain that concept to you, I must first describe how amino acid residues will be treated in it.

Any one of the amino acid residues, that is part of an active site, is unique in relation to the position of the other amino acid residues. In other words, it might be the same type of amino acid, as another residue in that active site, but it is not "the same" amino acid residue. Let me put it to you this way: one amino acid cannot occupy two positions on an active site- or on a protein for that matter- at the same time. So the first amino acid residue is one amino acid in the pattern of amino acid residues forming the active site, and the second amino acid residue although of the same type is distinct. It has its own unique position in the pattern of amino acid residues that make up the active site. The linear sequence is not the only thing to consider when calculating the odds of a three dimensional protein sequence. When the protein is folded, what also needs to be taken into consideration is the spaces to the left/right, above/below, and diagonally of each amino acid residue in the three dimensional conformation of the chain in order to calculate the odds of a particular active site being produced. In other words, any arrangement of amino acid residues, that form an active site, has countless other ways of rearranging themselves, sequentially, to make the same active site but just in a different three dimensional orientation.

I came up with a simple model to help visualize this concept. It helps illustrate the relationship between the three dimensional placement of the amino acid residues on the active site and their linear placement in the protein's sequence (3D-sequential-residue relationship). The active site of any protein must, by convention, have an empty space that fits the molecules it is trying to modify and envelope them in its amino acid residues. The variety of shapes that all proteins can take is mind boggling but they all have two things in common: one is the empty space in the middle and two a specific pattern of amino acid residues that surround that space. Regardless of what shape an active site takes, or how a protein forms it, these two commonalities will always be true. These commonalities are the key to all enzymatic activity, and to our ability to simplify all active sites of all catalytic proteins in order to construct a simple model; needed to explain the 3D-sequential-residue relationship visually.

This simple model consists of a multicolored ball that represents the active site, the colors of the ball represent the amino acid residues that form the active site, and a rope that represents how the rest of the protein forms that active site. The ball looks like a four colored beach ball split at the equator and shifted over by one color. So that instead of one section being one color from top to bottom, the top half would be one color and the bottom half would be another color forming an alternately colored, twelve sectioned pattern on the ball. The rope would wrap around the ball going over each colored section only once. Now let us imagine that the sections of color (amino acid residues) are attached to the rope (protein(s)) and that when we stretch out the rope the ball (the active site) dismantles into its sequential arrangement. Notice that the first color is red and that the colors following it correspond to the colors that form the pattern of the active region in this first orientation.

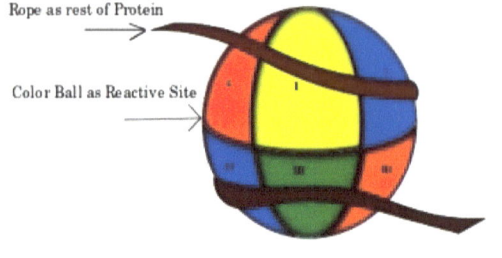

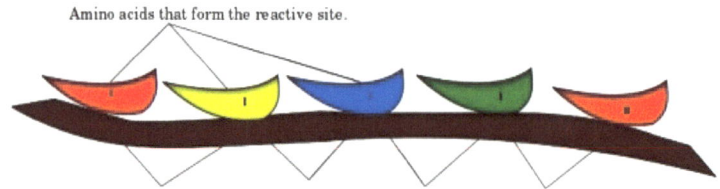

Figure 7. **This is how the Beach ball representation looks like when drawn out.**

Now let us take that ball (the active site) and turn it one full color (60°) to the left or clockwise when looking at it from above. Now let us, once again, stretch out the rope and see what is the new arrangement (sequence of the protein) of the colors now (figure 8). You will now see that the first color is not red but blue and that all of the colors have shifted down by one color in the sequence on the rope. As you can see, the pattern of the active site has not changed just its orientation; but with this new second orientation it has produced a completely new and different sequence.

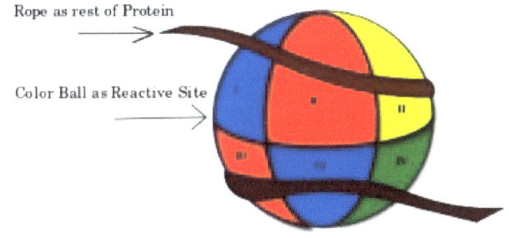

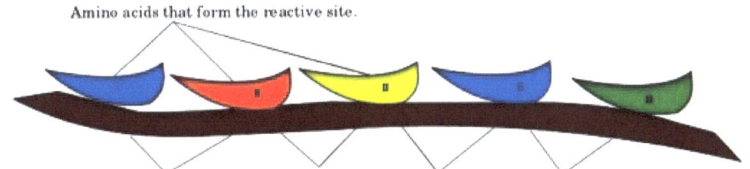

Figure 8. Now the Active site(Beach ball) has changed its orientation and see the change in the Amino Acid(color) order of the sequence.

Though the active site has a different orientation- due to the flexibility of the peptide bonds of the amino acids that make up the protein- it can still envelop its substrates and perform its enzymatic activity. A different orientation can also confer on to the protein the ability to envelop its substrates more easily and thus faster. This ability to re-orient the active site also explains why we find multiple versions of the same protein in the different kingdoms of Life; which have the same function but different sequences.

With any simple explanation there should always be a mathematical bases for that model to work. A mathematical expression that can be worked out and be used to either make predictions or show the high probability of something happening. In this case, we will be using a commonly known mathematical expression to show the number of sequences (different orientations) a single pattern of amino acids can have. We then take this number and compare it with the number of sequences a string of amino acids can have. Here is a hint of what we will always find; a three dimensional pattern will always have more sequences than a one dimensional pattern.

When looking at this simplified version of an active site, we need to remember that we are dealing with a three dimensional sphere. So once we pick a starting point on that sphere, whatever amino acid residue was used there, the rest of the amino acids must recreate the original pattern around it. This creates one distinct sequence, from millions, that this protein can have in order to form the same active site, but with a different orientation. Now if we take that same point on the sphere, but start reconstructing the active site with a different amino acid from the same pattern, again, we are going to end up with a different sequence that still recreates the same active site but with another distinct orientation.

Every point on that sphere has at least eight places, i.e. orientations, around it to put the next amino acid residue of that pattern. With every point on that sphere having at least eight orientations to recreate the active site we could believe that there are limitless ways to sequence the pattern. The reality is that depending on the size of the active site we get the number of amino acid residues and that gives us the length of the sequence. That being so, it is still a stagering number of sequences at any length. So then how could we calculate the number of sequences for all the different orientations that an active site, of any length of amino acids, could have?

Thinking about the number of points on a sphere and the number of orientations around each of those points can be very confusing for anyone trying to solve this problem. But, what we must do, in order to solve this problem, is do what I have always done with problems in electronics or nursing: I come at them from the opposite direction or turn the problem upside-down. In this case, instead of calculating how many patterns, we calculate how many different sequences can be made from the number of amino acids that make up an active site. It is like this, once you pick an amino acid to start a pattern the rest of the amino acids that come after it have one less way of orienting themselves.

In mathematics when the order of a sequences of numbers is important you perform a factorial. As you select one number to multiply by you have one less the number of choices for the next multiple. Like if you have 5 items, you put one down, you now have 4 items to choose from for the next position. Now to calculate how many different patterns of items you could make, you perform a factorial of the number 5. The formula for factorials is this $5! = 5 \times 4 \times 3 \times 2 \times 1 = 120$. So you would have 120 different ways of lining up your items. So how does this solve our three dimensional to sequential sequence conversion problem?

Even though, with a sphere, we have limitless ways of orienting any pattern in three dimensional space, we only have a limited number of ways of doing it sequentially. As we string together amino acid residues into a sequence, every time we add one from the pattern we have one less amino acid; hence one less way of orienting that pattern on the sphere. So regardless of the identity of the individual pieces of a pattern, you only need to know the total number of pieces in order to figure out how many different patterns you can make. Although, there are only twenty amino acids that Life uses to make every protein it needs; it needs a lot more than twenty to make each particular pattern. Life also uses different patterns to make different proteins that do the same job. What seems to be most consistent with these different proteins, that do the same job, is the length of the segment that forms the reactive site (Darst et All.)

So using this principle let us apply it to the multicolored ball example. The ball has twelve sections, each a single color. There are only four colors to choose from for each section. Remember, in this example each section and color represents one amino acid residue. So, if there are twelve sections and each section can have four different colors, and we take the rope that forms that beach ball and lay it out flat; we can calculate how many different sequential patterns can be formed by calculating: $4^{12} = 16,777,216$. Now here is where it gets exciting: If we take one pattern, that forms a useful reactive site, and calculate how many different orientations it could have in three dimensional space; we would accomplish this by performing a factorial of the number of sections (amino acids) that form the reactive site like this: 12! = 12 x 11 x 10 x 9 x 8 x 7 x 6 x 5 x 4 x 3 x 2 x 1 = 479,001,600. This much larger number makes it abundantly clear that the three dimensional pattern has many more opportunities to form the useful reactive site than the sequential pattern has sequences.

Now you may ask: How is this possible? Well if we look at the sequential method: It only uses four colors on each position; which does not take into account the overall pattern. On the other hand: when we take into account the overall pattern, we realize that even though the same color might occupy the same space; it might not be the exact same color from the pattern; the red in the space now is the second appearance of red in the pattern; hence a unique red that needs to be accounted for, too. When we choose a starting space, at first we have all available colors and their re-iterations in the original pattern to put in this space. Then in the second space, next to the first, we have one less color to choose from, but the main thing to consider here is: that on any space on a sphere all of the available colors can be used depending on the orientation of the active site; and coming at it from the other direction, depending on which two amino acids we start with determines the orientation of the entire active site. All these possibilities multiplied by all these spaces on a sphere can give us the millions of patterns, and orientations, predicted by our factorial formula.

What all of this basically boils down to is this: although there might be, though I highly doubt it, only one pattern out of 16 million (using the beach ball example) for amino acids to form an active site that performs one particular reaction; Life has 479 million different ways of stumbling onto just that one pattern! Now there are several hundred million ways for life to stumble onto just this one pattern for this one reaction. What about all the other millions of reactions that Life is capable of performing? I am sure that in the rest of the other millions of patterns- that are left- all the other reactions are contained. And for each one of these patterns there are also hundreds of millions of ways for Life to stumble onto each one of them too.

These numbers are just for the patterns of amino acids that make up the reactive site. What about the rest of the amino acids that form the complete protein? Again we have something that has long been known but since was not properly applied to the analysis of protein structures. The majority of all the different types of amino acids when strung together form both alpha-helixes and beta-ribbons spontaneously depending on the certain pattern of amino acids. Although the patterns of amino acids that form these shapes are limited. They do not have to start or stop with the same amino acids; therefore the exact sequence of amino acids that can form those shapes are literally unlimited. What is most important, for the functioning of the enzyme, is not the exact sequence of amino acids but the final shape of the rest of the protein. What this means for the formation of the rest of the protein, of an important reactive site, is that there are literally trillions upon trillions of other sequences that would form the exact same shape. So, to get the odds of an important enzyme forming can be arrived at by multiplying the trillions of sequences that can possibly take the shape of the protein times the hundreds of millions of sequences that can mimic the reactive pattern times the millions of different reactive patterns.

So it looks like, to me, that in three dimensional space it is more improbable for Life not to stumble upon a useful amino acid pattern; one that would create a reactive site; which would then perform a useful reaction for Life; therefore Life in this universe is inevitable! But another amazing thing we have proven here is that any universe, capable of producing carbon atoms in three dimensional space, would always produce life.

Up to this point, in this chapter, I have also been expounding on my massive conjecture of, what I now call, the Trisapient Homolog Process (which I am patenting). The Trisapient Homolog Process is the process by which a known sequence of amino acids that form an active site can be re-sequenced to form the same active site but in a different orientation. Now this different orientation could cause the enzyme to react more quickly; if it is in a more favorable position or more slowly if it is not. This would allow man to redesign existing enzymes in order to control their reaction rates.

My Theory posits multiple enzymes that synthesize RNA. So let us take a look at the enzymes that synthesize RNA today. All the types of cells and organelles: prokaryotic, chloroplast, archaeabacteria, and eukaryotes, have their own uniquely-sequenced enzymes that synthesize RNA. But what they do have in common are two alpha-helixes and two beta-sheet structures that are all identical in structure. The alpha-helix is the most common structure that amino acid sequences form in proteins. In the article Conformational flexibility of bacterial RNA polymerase. Darst et All. compared the amino acid sequences of the E. coli RNApolymerase enzyme with one from the Thermus Aquaticus bacteria. What they found was that 58% of the sequences were similar. Not "exact" but similar, but let us just concentrate on the 42% portion that were not similar. That is a very large degree of difference especially when considering that just one wrong amino acid causes sickle cell anemia. It is amazing that such a large difference in amino acid sequence still causes the same biological activity; the Trisapient Homolog Process gives a very probable explanation for this and appears to be the best interpretation of the evidence.

This was, for many years, the best evidence I had to support my Trisapient Homolog Process. I knew that I had to somehow show that the amino acids in these different proteins had the same spacial arrangement, in their respective active sites, in order to prove that the Trisapient Homolog Process is true. Again, instead of trying to reinvent the wheel and trying to figure out the amino acid sequence of these proteins in three dimensional space and comparing them; I decided to do an online search for any similar experiments that might have already been done. And after- only thirty minutes of- searching through the results for the words: homolog amino acid sequence; I came across a paper with results that perfectly confirm the Trisapient Homolog Process.

The name of the article is NRMT2 is an N-terminal monomethylase that primes for its homologue NRMT1, by J. Petkowski et all. The article states that the protein NRMT2 (N-terminal regulator of chromatin condensation 1 methyltransferase 2) was for a long time called METTL11B (methyltransferase-like protein 11B) but since it acts on the same substrates as NRMT (now renamed NRMT1) it was given its new name. It also states that NRMT2 has sixty more amino acid residues than NRMT1 and if you remove those extra amino acids their sequences are still only 75% similar not exact. There have been many articles that describe how dissimilar homologues are. What makes this article special is the following: it shows that the amino acids that form their reactive sites occupy different locations on their respective sequences (different residue numbers in figure 9), that their reactive sites have different orientations, but ,and this is the key proof, the reactive amino acids-that make up the reactive sites- are in the same relative positions. So although their sequences are different they create the same reactive site just in a different area of and orientation on their respective proteins. Figure 9 and its explanation are both excerpts from the article. In the picture the proteins are superimposed to show the amino acids in the same positions but just in a slightly different orientation over all; a perfect visual proof of the Trisapient Homologue Process.

Not all combinations predicted by probability would be produced or would be favorable. By contrast biologically active combinations we know would be more favorable under the conditions for ECE; we know this because that is what we see today within cells. Being that the most abundant structure we find in biologically active proteins today is the alpha-helix, it is not to far of a leap to assume that alpha-helixes would have been the most commonly formed prebiotic proteins. Why should we expect alpha-helixes to dominate in a prebiotic environment? The same reason it is the number one secondary structure of proteins today: all amino acids, except proline and glycine, readily form alpha-helixes. They spontaneously take this conformation without the aid of other proteins or chemicals; thus not surprising to

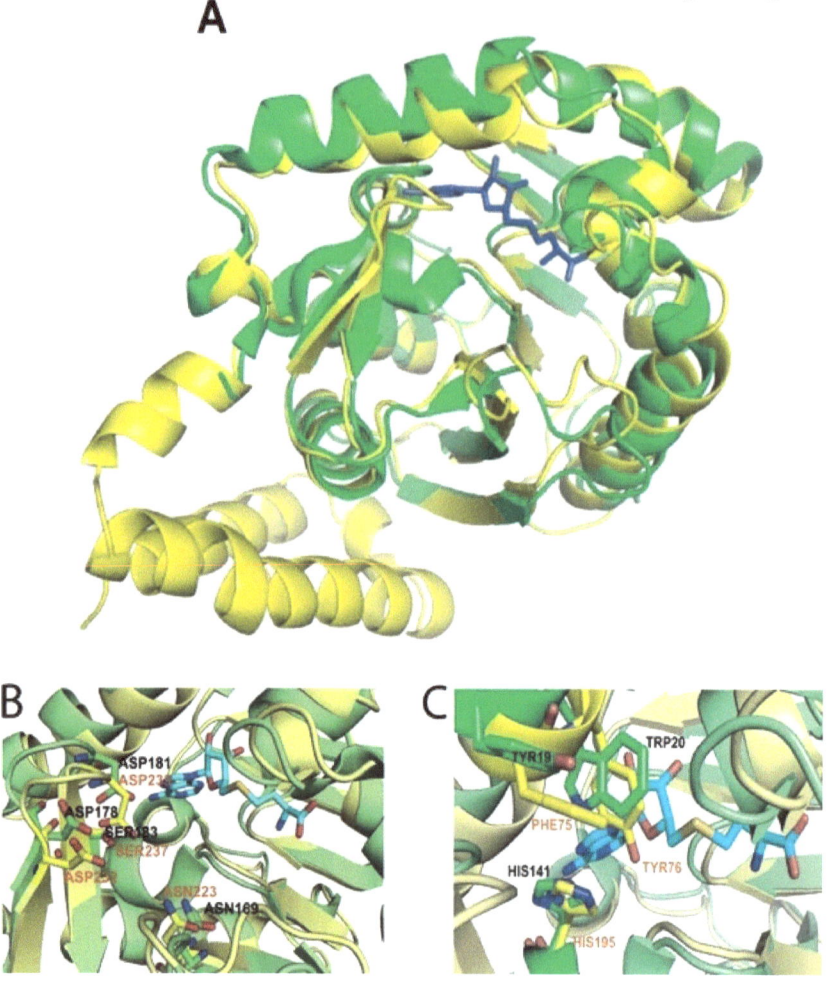

Figure 9.(A) Homology-based model (as calculated by the Robetta server) of NRMT2 molecule (yellow) having the same overall fold as NRMT1 crystal structure (green; PDB code 2EX4). The position of the S -adenosylhomocysteine molecule in the NRMT1 crystal structure is shown in blue. The N-terminal 60-amino-acid domain of NRMT2, which is not present in NRMT1, is modelled as two helices. (B) Critical catalytic residues in NRMT2 (orange font) occupy the same spatial positions as in the NRMT1 crystal structure (black font). (C) Conserved aromatic resides in NRMT1 and NRMT2 form a chromodomain-like arrangement of aromatic residues and may be responsible for binding methylated substrates and products (J. Petkowski et all 2013)

find that all of the important proteins in Life today have this structure. This also explains the large number of amino acid differences in the sequences of homolog proteins. Since large portions of the protein are not part of the reactive site they are free to take on any sequence that still makes an alpha-helix. This is why we see up to a 40% difference in the amino acid sequence of homolog proteins but they can still carry out the same exact chemical reaction; this also applies to beta-sheet conformations, too.

The Trisapient Homolog Process (THP) would have undoubtedly produced an unaccountable amount of different small proteins. And given the nature of proteins, with their propensity for forming ionic bonds and van der wall forces, they would have undoubtedly been interacting with each other. Interacting in ways that would have caused the small proteins to unite with each other and make larger proteins. Much like we see in modern cells today. In modern cells, they encode in their DNA the information for making several small individual proteins. These small proteins would then unite to make a larger protein. All that work just to perform one specific chemical reaction. Individually, these small proteins cannot perform any chemical reactions until they unite with their counterparts creating a collective reactive center. I do not see why we would expect it to be any different when THP was in full force back then. Especially when, for over 600 million years, meteors and asteroids impacted the amino acid laden oceans, which modern experiments confirm, would have made trillions of polypeptides with every impact.

The most important chemical reaction, for Emergent Chemical Evolution (ECE), that would have been peformed by small proteins uniting is the catalysis of the first step in the pyrimidine biosynthesis by the enzyme Aspartate Carbamoyltransferase (ATCase).This is also the first step in the synthesis of RNA. ATCase, in Escheria coli, is a 12 unit protein complex. It is made of 12 small individually encoded proteins that are bound together by ionic and hydrophobic interactions between their amino acid residues. Meaning that these 12 smaller proteins self assemble into a larger enzyme. Six of the subunits are catalytic who together make the reactive pocket using only 12 amino acids that perform the catalytic reaction (I came up with my beach ball example long before discovering this information) and the other six subunits are feedback control units (Ke et al., 1984). In contrast more complex bacteria and animal cells use large single unit proteins to perform the same reaction. Why would early life, which E. coli is a perfect example not having mitochondria, would have encoded such an important enzyme in such small pieces instead of one large enzyme? In order for E. coli to have gained an evolutionary advantage from genetically encoding only one small portion of this enzyme complex at a time; the other pieces must have been easily available in the environment being continuously created by ECE.

ECE through THP would have been producing many types of polypeptides of every length, uniting, non-uniting, globular shaped, sheet shaped, catalyzing, non-catalyzing, and in many cases, as seen today, in many cofactor roles. The variety of ways that life carries on its chemical reactions with, and without, cofactors should have made researchers wonder; why does Life perform the same reaction in so many different ways? To me the easiest solution is- what I have stumbled upon- ECE, through THP. THP would have provided the necessary components for whatever reactions Life needed. Of course the individual organisms, through random mutation, and with the probability predicted by THP would have accumulated the genetic information to reproduce some of those components quickly and independently. Whenever that would have happened, it would have given those organisms an evolutionary advantage over their competitors. They gained that advantage by being able to perform those reactions so much quicker. Quicker in a real life and death game of molecular musical chairs; where those that grab the needed molecules quicker survive.

ECE with the probability predicted by THP would have easily produced a six unit catalytic only homolog version of ATCase (having 479 million ways of arranging the 12 amino acids that make up the reactive pocket).

The way the protein subunits are arranged, the control ones being on the outer surfaces, tells me that those evolved later. The control proteins use allosteric control which changes the conformation of the inner catalytic enzymes. Whether ATCase evolved susceptibility to these controls, or can only form with allosteric controls built in, or needs the conformational changes to take place in order for it to catalyze its reaction the ERCs of ECE would have readily served either one of these purposes. All proteins' conformations are effected by the pH of its environment. The ERCs would have taken the ATCase homolog through the pH gradients which would have made the enzyme catalyze its reactions without the need for any allosteric control subunits.

In modern life allosteric control in most cases is activated by ATP. When ATP attaches to a binding site on an enzyme it causes a conformational change in that enzyme. That conformational change causes the enzyme to catalyze its reaction. So, what life has engineered with the synthesis of ATP is the storage of mitochondrial pH gradient energy, and what ATP does is deliver that energy by remotely activating a conformational change- the same change a pH gradient would have had on an enzyme- when it binds to an enzyme's activation site. So before the emergence of Life the Emergent Recursive Cycles (ERCs) would have played the part of ATP by taking the enzymes through the pH gradients; thus causing conformational changes in them making it possible for the enzymes to catalyze their reactions; thus powering the emergence of Life.

The second step in the synthesis of RNA is performed by glycinamide ribonucliotide synthetase (GARS) which in E. coli is an enzyme composed of 429 residues of which 193 residues are dedicated to the binding of ATP. These ATP binding residues would have been unnecessary during the time of ECE, thanks to ERCs, leaving us with an enzyme that only needs to be 236 residues long. The E. coli version of this enzyme has only between 41 and 51% similarity with the same enzymes from: Bacillus subtilis, a bacteria; Saccharomyces cerevisiae, a fungus; and Drosophila melanogaster, a fly. GARS in humans is part of a much larger trifunctional protein who's later appearance in evolution and sequence difference from lower organisms is easily explained by THP. Such a large degree of difference- not only between organisms from other kingdoms of Life but surprisingly from other bacterias as well- tells me not only that THP was continually used by Life; but that, in contrast, it could have easily brought about a GARS like enzyme at the beginning of Life. Clearly THP is not only responsible for the appearance of homologs of important proteins in modern Life but also in their appearance before and the cause of life's emergence.

Here I will show you another example where THP is a better explanation for what is found in the homologs of another protein necessary for the de novo production of RNA. Grande-Garcia et al in their article Structure, Functional Characterization,and Evolution of the Dihydroorotase Domain of Human CAD compared the amino acid sequences of Dihydroorotase enzymes from human, E. coli, and seven other organisms that are representatives of bacteria, fungi, thermophile, and archaea to determine the structure of human DHOase (huDHOase). In one figure of the article all of the sequences from the nine different organisms are stacked one atop of each other. They are aligned so that the amino acids that touch the substrate (ie the amino acids that form the active site) form a straight line from top to bottom. When this is done an amazing thing happens. All of the amino acids- that form the active site- in all nine organisms are the exactly the same amino acid. Now here is the problem, 70% of the amino acids that make up the rest of the enzymes do not match at all. Not only that but in the sequences belonging to the lower organisms, there were many large gaps in their sequences due to their enzymes being smaller compared to that of human's.

The number and size of these gaps as well as the extreme differences in their sequences pose a serious problem to the idea that these enzymes evolved from a common ancestor. Take the E. coli's DHOase enzyme which has the most and largest gaps compared to all the other organisms. Then take a human's and superimpose their active site amino acids in three dimensional space which Grande-Garcia et al did in their article Structure, Functional Characterization,and Evolution of the Dihydroorotase Domain of Human CAD (Grande-Garcia et al 2014). When this is done you can clearly see that the same amino acids, once the active sites are aligned, occupy the same positions in three dimensional space (see figure 10). Once again this is another stunning confirmation of the validity of THP and its ability to produce homologs of important enzymes throughout time and the domains of life.

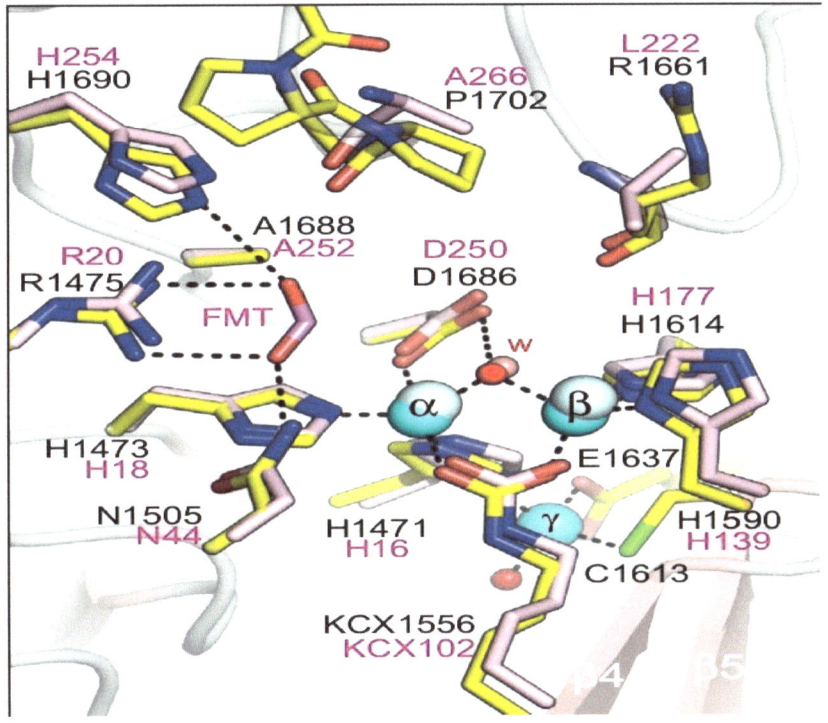

Figure 10. The superimposition of the apo forms of huDHOase (carbons colored in yellow) and ecDHOase (carbons colored in pink) is a perfect example of the veracity of THP. Zn 2+ ions and water molecules are represented as cyan and red spheres, respectively. In huDHOase, a formate molecule from the crystallization condition mimics the interactions of the substrate a-carboxylate thereby completing the exact pattern.

Surprisingly this exact match of amino acids in 3D space does not bode well for the proposition that these enzymes evolved from a common ancestor. First of all the DHOase enzymes from these organisms have distinct organizations; where the helixes of the alpha-helixes, that form the active site, have different orientations and curve directions. So although the position of the reactive amino acids fall in the same positions in 3D space (necessary for catalysis to happen) they arrived at this by completely different strategies. This is exactly what THP predicts would happen. It is also very interesting to note that the number of active amino acids in this enzyme is also twelve (giving it 473 million different orientations to form the same reactive center). Being that these, amino acids, are in the exact spots needed for catalytic activity, any change in their sequence number (making them longer or shorter) would have changed their location rendering the enzyme ineffective. That is why these, enzymes, could not have evolved from a common ancestor and instead, as predicted by THP, would have emerged independently. Independently as a result of random addition and random duplication mutations of non vital proteins.

When looking at the most important proteins in Life, we see the most easily formed structures dominating; should this not have been a big clue for us to see how Life may have gotten started. We do not find very complex structures that need special conditions to form their shape; they form spontaneously. We do not find strange chemicals that can- only- be made by Life being assembled into proteins; instead we find the most abundant organic compounds- amino acids. Which can be found in the most common meteorites. That are easily made from any mixture of elements. When united, need no help forming alpha-helixes or beta-sheets the two most common shapes in all enzymes. Apparently amino acids seem to be a better candidate for initiating the reactions that started Life on Earth; more so than any other biopolymer: ie RNA, DNA, or the byproducts of the kreb cycle. Emergent Chemical Evolution using amino acids, in accordance with the Trisapient Homologue Process, form a mechanism that repeatedly recreated the reactions needed to initiate Life without the need of any information storage or any translation mechanism!

19 FROM KINETIC ENERGY TO CHEMICAL ENERGY

Emergent Recursive Cycles (ERCs) were the driving force that provided the energy to power ECE. The means by which the kinetic energy of the currents could have powered the chemical reactions in ECE is not immediately apparent. We, as human beings, experience the world on a macro scale and can easily distinguish between kinetic energy from chemical energy. We now know that we are powered by chemical energy derived from the food we eat. At the molecular level though it is not so easy to distinguish between what is chemical energy from what is kinetic energy. At the macro scale kinetic energy can be easily identified. You just look for an object that is moving relative to its surroundings. At the molecular level it is not so easy to determine what is kinetic or chemical action.

    As the currents transported these polypeptides across the gradients- created by the layers of different pH levels- their ends became chemically active. Hence, the kinetic energy of the currents was converted into chemical energy. The chemical energy necessary to make these polypeptides bond with each other, amino acids, and/or metal ions. And as the currents were predominantly powered by the rotation of the earth- which in turn is powered by gravity- I, here by, make the fallowing assertion: that gravity was what powered the initiation of life. I also posit that because of this no solar energy was necessary at the initiation of life. Tadda!!! Again my theory solves another problem that the others cannot. The problem of the sun's low energy output at the time of life's initiation is very easily explained by Emergent Chemical Evolution.

It is gravity that holds the water against the Earth's surface to create the oceans. It is gravity that causes the Earth to rotate which in turn creates the relentless currents in the oceans. It is gravity that holds and presses the atmosphere to the Earth and its oceans causing the pH gradients. It was gravity's unending force that provided the limitless kinetic energy that powered Emergent Chemical Evolution when the Sun's output was too low to do so.

Once gravity initiated ECE and helped it to produce countless proteins and therefore enzymes of all types it did not just stop providing Life with energy. Gravity would continue to do what it does best provide limitless kinetic energy by taking these proteins/enzymes relentlessly through the oceans' pH gradients. As these proteins/enzymes passed through the pH gradients they would become denatured; making it easier for them to envelope their substrates. As the enzyme-substrate complexes (ligand enzymes according to biochemical parlance) approached a pH level that allowed the enzyme to resume its natural, low energy, state; the enzyme would then react with its substrates. This is the primary way that the kinetic energy of the currents powered the mechanism of Emergent Chemical Evolution.

Remember that ECE has several hundred million different ways of stumbling onto every one of the millions of different patterns of amino acids that can create a useful enzyme. Eventually several of these enzymes would be able to produce adenosine triphosphate (ATP) but these enzymes would be much smaller than their modern day counterparts. They would lack the structures that would anchor them to the cell membrane making them much smaller thus easier and quicker for them to emerge from ECE. But these smaller proteins would work even without having membranes like their modern day counterparts.

The following will explain why those enzymes back then would have worked without a membrane. The Kreb's cycle uses one molecule of glucose to make 36-38 ATPs, surprisingly it only makes 4 ATPs directly from the glucose molecule, but the rest of the glucose's energy is used to create a hydrogen ion concentration gradient across the mitochondria's membrane. It is the movement of the hydrogen ions through the enzyme embedded in the membrane that makes the rest of the 32-34 ATPs; hence why Emergent Chemical Evolution's mechanism of the currents transporting enzymes through the ocean's pH gradients exactly mimics the overall action of the kreb's cycle.

Instead of the hydrogen ions moving through the enzyme as a result of a high concentration of ions across a membrane, the currents would have carried the enzymes through the pH gradients in the water instead. This would have caused the hydrogen ions to constantly pass through the enzymes continuously charging them and producing ATP. That these ATP enzymes were floating about, freely, would not have been a problem; because they would have become energized, and then later pass the electrons to adenosine diphosphate (ADP) selectively. In modern cells, mixed with all kinds of other molecules, these types of enzymes do not lose their electrons until ADP mates with them as well. So the concept that these enzymes were floating around and still able to work is not as far fetched as, I am sure, some would like to think.

Another reason, ECE is more likely to have been the initiator of Life, is because all of the molecules used in the kreb's cycle are derived, in modern cells, from amino acids. Amino acids, and here I find it appropriate to remind you, are the easiest molecules to be produced by physical processes, are ubiquitous in the universe, and are still what life reduces its nutrients to. At this point proponents of the metabolism first hypothesis would probably jump up and say "Aha Life started with metabolism first!" But they would be wrong. I only bring it up here because of the subject of this chapter. Later in the book I will describe how kreb's cycle evolved.

What I am describing here is a natural and

deterministic way for chemical evolution to occur and be powered. The kind of chemical evolution that modern scientist are creating in their labs today in order to come up with novel proteins; of course, minus the scientist. His part is played by the ERCs and the parts played by the laboratory equipment are played by the oceans and the carbon dioxide laden atmosphere. Then every completed ERC through the upper layer of the oceans represents the equivalent of one experimental run of these modern experiments.

Please do not be mistaken- the experiments in artificial chemical evolution, I read about, were not the inspiration for my theory. The reason I point out these experiments is to show the feasibility of this kind of chemical evolution. I must also point out that these modern experiments yielded desired outcomes rather quickly. Which is the result of the Trisapient Homolog Process we figured out and calculated in the previous chapter. Now put all of these things into the perspective of ECE and add a geological time frame- and it quickly becomes very clear to see how easily life would have emerged.

20 THE FIRST LIFE FORM

Proponents of the other theories could now say, with the emergent recursive cycles (ERCs) in place assembling enzymes and those enzymes producing all types of molecules- such as sugars, lipids, etc.- that "Everything is in place for the emergence of the first protocell." And to them I would simply say "You are mistaken." The conditions for the formation of a protocell will never come. And that also goes for a self-replicating molecule for the many reasons given earlier in this book. Although modern cell membranes contain lipids, membranes are not solely made up of lipids. What makes the membrane work are the proteins embed in it.

Sure Emergent Chemical Evolution (ECE) produces all types of proteins, lipids, and every conceivable molecule by means of enzymes. And sure it is conceivable that a protocell could spontaneously form. But that leaves us with an even more difficult problem to solve. How-once the protocell is formed- could the information to make the protocell be encoded into the RNA? The actual distance between the protocell membrane and the RNA is very tiny, but- conceptually- it might as well be the distance across the Grand Canyon. What mechanism could have possibly reverse engineered and encoded into the RNA the information needed to make the proteins located on the already formed membrane?

This is the biggest problem with protocells. How did

they store the information to generate their structure, metabolism, and their method of replication? It is hard to imagine a mechanism that could have reverse engineered and then encoded into the RNA the information necessary to do what physical forces spontaneously assembled. Especially when the scientific community has spent over forty years conducting experiments trying unsuccessfully to replicate these exact scenarios; unable to make it happen even in the most perfect laboratory conditions. The only assumption that researchers have made until now that is correct is that Life as we know it did start with RNA. Though, not by means of self-replication or catalyzing reactions always performed by proteins today.

At this point, after having shown how large enzymes can form from smaller polypeptides; how easily with THP they can form important enzymes; how these enzymes could be powered by ERCs through pH gradients; and how THP can be shown working independently throughout time and the kingdoms of Life. We can now say that a series of enzymes would have easily emerged in the prebiotic oceans that would have led to the production of RNA nucleotides and their sequences. Although the enzymes used to make RNA in all lifeforms are not all the same, the sequence of reactions and molecules produced by them at each step are exactly the same throughout all the kingdoms of Life. Meaning that if all those different sequences still produce enzymes that make the exact same molecules; THP is the best explanation for this. All that being said, I think someone would be hard pressed to level the charge of hand-waving RNA's appearance on me at this point. At this point we have a larger problem. Even with ECE producing all those proteins and, due to that, we have the eventual appearance of RNA. The problems with the spontaneous formation of protocells still holds.

While I was mulling over this very problem in my head the answer occurred to me. It occurred to me, after asking myself a series of questions about how life works today. What does an organism's genome encode for? Proteins. What organism has the fewest genes? Can this organism of fewest genes self-assemble? Is there an organism that encases itself only in protein? When I, finally, came up with the answer I was on my way to nursing school. The answer nearly caused me to hit the car in the next lane! The answer was a Virus!

I remembered that viruses are RNA and DNA strands encapsulated in protein coats. I then remembered that the smallest of the viruses has only four genes, that they replicate and mutate exponentially, and then I also remembered in astonishment that the tobacco mosaic virus (TMV)- a virus of only four genes- self-assembles (Butler and Klug 1978). Eureka!!!

A virus, as the first form of life, makes perfect sense when viewed from the perspective of Emergent Chemical Evolution. They replicate inside a cell where the conditions are the same as those conducive to ECE. They do not have mechanisms to replicate themselves; they have to hijack the cells. Which in that case, ECE makes a much simpler explanation for the origin of viruses. Because I cannot conceive of a mechanism that can cause DNA- that was already encoded to make a cell- to spontaneously start assembling viruses; which is the current model for the origin of viruses.

The virus as the very First Life Form makes perfect sense. Well technically, they are still debating whether viruses are alive or not. I am here to tell you that the debate is over. They are both alive and dead. Let me explain how they are both alive and dead. As we have seen, in the evolution of species, it takes a series of small steps to go from one species to another. So why should it be any different going from non-living chemicals to life? A virus fits perfectly in the space leftover from non-living molecules to living cells. Separately the virus's individual constituents are not alive- but neither are the cells- though they do have one unique quality; when you put them in water they self-assemble (something a cell cannot do) into a life form- The Very First Life Form- a Virus.

I know that under current definitions of life the virus does not qualify. The virus does not grow. The virus does not consume and metabolize food. But it does gather material from its environment- whether within a cell or an acidic body of water- and replicates exponentially. It does- by replication errors- mutate and thus evolve. In order to qualify a virus as a living thing we need to be a little more flexible with the definition of what is living.

The definition of life was conceived when our understanding of its function and history was limited. I feel- just like everything else in science- we should review and change, at the very least redefine, our positions and definitions as we gather more information. I feel that the definition of life has long been overdue such a revision. What I am proposing is not a radical redefinition of life; but instead just a little refinement of the definition based on more recent information.

One of the refinements I am proposing is to delete the requirement for life to grow- which only applies to multi-cellular organisms anyway. This rule was made before we realized that for 80% of the history of life there was only uni-cellular life and that, for the most part, the individual cells did not grow in size. Another refinement I propose is to change "consumes and metabolizes material from" to "incorporates material from" the environment in order for it to replicate. Another is to change the rule from "it must self replicate" to just "it must replicate."

I know that some of you are going to accuse me of just changing the rules to those that suit my theory, and I would have to agree with you. In order for me to come up with my theory it was necessary to view things from a molecular perspective. Using this perspective I cannot help but see consumption and metabolism just as another more complex- i.e. more evolved- way for an organism to gain useful molecules from its environment.

I also think that looking at replication from the molecular perspective makes it necessary to change the replication rule. What I see-in cells- is an organism changing its internal environment to the conditions necessary for ECE in order for its internal molecules to replicate themselves. When Life was at its viral stage it did not need to recreate these conditions because they were already present; hence why they could replicate without cells. Besides, no organism replicates in a vacuum. They all must consume extra material from their environment in order to replicate. Thus, with this refined definition of life, we can now look at viruses not only as alive but also as the simplest-i.e. Least Evolved- form of life.

21 DUMB RNA

As I see it, Emergent Chemical Evolution (ECE) is an eloquent solution to the problem of which came first: metabolism, proteins, or RNA. Clearly proteins were the first products of ECE, and this was not only a simpler way for proteins to form but for RNA as well. Now with both proteins and RNA on the scene it is not hard to imagine that at some point a virus would have self-assembled just like they do today. Even if this was the case, which it is not, it still leaves a large gap between the self-assembly of a virus and the complex mechanisms that make a living cell possible. The first thing we need to establish is what is the minimum set of RNA nucleotides necessary to self-assemble a virus. In order to accomplish this we need to identify the simplest virus. One that has the least protein components and least number of genes encoded in its RNA.

For a while I was under the mistaken impression that the simplest virus was the tobacco mosaic virus (TMV). It is composed of only RNA and one type of protein that self-assembles and encapsulates the RNA. TMV has only four genes in its genome. Reading the article that described TMV's self-assembly was very important none the less. In that article it described how the proteins that encapsulated the RNA would lose their shape when the pH of the water would change. This change in shape would cause the proteins to disassemble and release the

RNA for replication. The authors of the article stated that the most important variable in the self-assembly of the virus was the "pH" of its liquid environment (Jonathan and others 1978). When it comes to activating an enzyme to breakdown a protein, a hemoglobin to release oxygen, mitochondria to produce ATP, ribosomes to assemble, or the replication of DNA and RNA a change in pH seems to be the central chemo-mechanical force that actuates all of Life's reactions.

The search for the simplest virus, one with the least amount of genes, led me to a type of virus I did not know even existed, the viroid. The viroids are a very special kind of infective RNA molecule; that is correct it is a virus that lacks a protein coat altogether. The fact that it could propagate without a protective protein coat was curious enough, but what made it even more so was that its RNA contains no genes! These viroids encode no proteins or enzymes of any type (hence the name viroid). They just find their way to the nucleus of a plant cell, and simply, or should I say aggressively, take over the RNA replication system and make copies of itself. The viroids are RNA of only 375 to 390 nucleotides long. Apparently this is the ideal length for RNA to go through conformational changes that drive replication of itself. This length and conformation are so ideal for replication that without the aid of its own proteins it takes over the RNA replication system of the cell it invades. A very useful emergent property of such short lengths of naked RNA.

Once I noticed the similarities between tRNAs and viroids, I predicted that viroids would take a circular conformation and replicate by rolling-circle mechanism. When I looked for evidence of this online I found an article called Viroid Replication: Rolling-Circles, Enzymes and Ribozymes (Flores et al. 2009) that proved my assumption right. Again another piece of modern peer-reviewed and published evidence in support of ECE; written by a world renown expert on viroids.

What we have in viroids are relics of the first replicating molecules. Notice I did not say self-replicating molecules. We have clearly established that ECE would have easily produced enzymes capable of producing RNA nucleotides and RNA sequences. These RNA sequences would have been a bunch of randomly joined nucleotides; not readily coding for any proteins. They could not code for them, even by accident, because a system of translation had not yet emerged, but what they did do was replicate in conditions conducive to ECE (like inside a cell). Viroids are exactly the type of random RNA molecules that we would expect ECE to produce.

So the first Life forms are not only trillions of viruses but stripped down versions of them. Pieces of Dumb RNA. Dumb because they do not contain any information for producing proteins, and do not have the ability to replicate themselves by themselves. Dumb because they do not direct their own activity. They are victims of the ERCs taking them through the pH gradients. Their activity is then dictated by what level of the pH gradient they find themselves in.

At one level of pH they might replicate , at another they might splice themselves which will form more tRNAs, and yet at another level of pH they will take on twisted conformations protecting themselves from attack or degradation. Being that these viroids would replicate exponentially just like they do today. It would not be unreasonable to assume that with, almost, every replication errors would be introduced. If modern viruses replicate exponentially and mutate exponentially due to all the errors introduced during replication. It would be especially true back then, too. Easily, every type of tRNA would have been produced and autocleaved from the viroids within a very short period of time

It is not hard to understand why viroids are shaped the way they are. Back at the beginning when RNA was being combined into viroids they were not being copied by rigid proteins; like modern viruses who use RNA-Dependent RNA Polymerase. So since the RNA was replicating by a rolling-circle process with loosely held together proteins the RNA would attach haphazardly creating the distinctive shape of all the viroids. This action would also aid in the appearance of RNA strands that would act as amino-acid-attaching-to-tRNA enzymes.

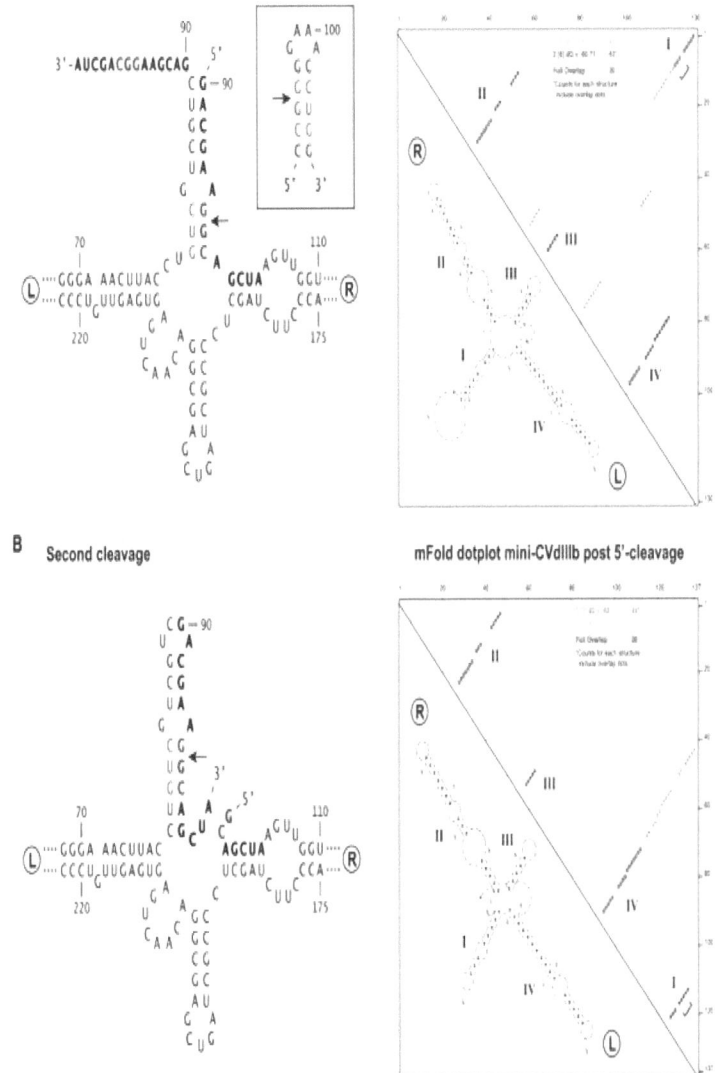

Figure 11. A Viroid's mid section has the same shape and number of nucleotides as those of many Transfer RNAs who's nucleotides number from about 70 to 90.

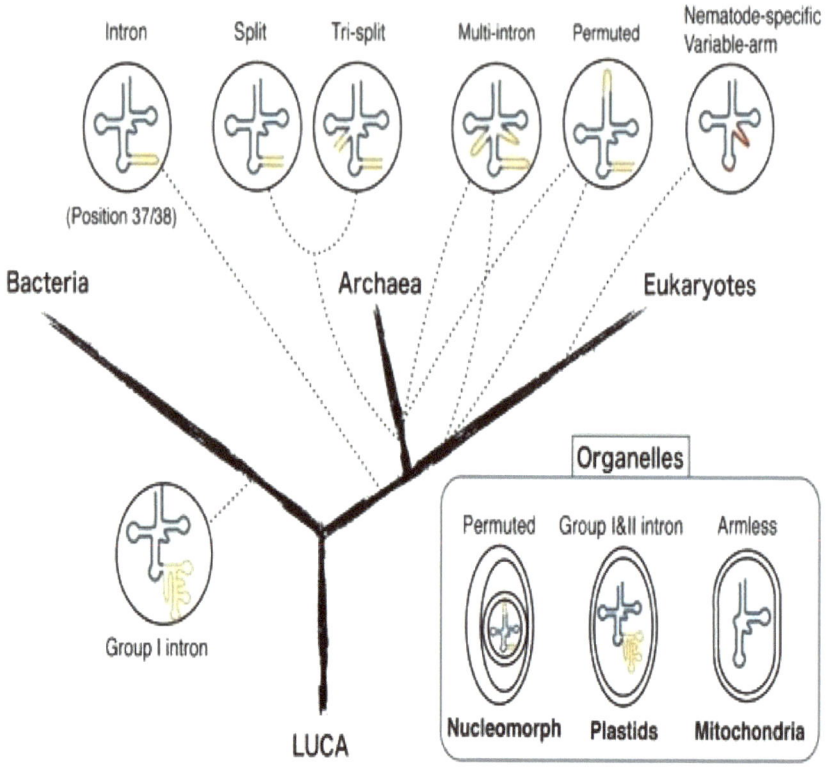

Figure 12. These tRNAs have introns that are autocleaved from the same sites that viroids start replication and also have further attachments. These introns give further evidence that tRNAs were the byproducts of viroids autocleaving themselves in the prebiotic oceans.

In fact in modern Escheria coli there is an enzyme made of some proteins and 375 RNA nucleotides (Funny how its the same length of nucleotides as a viroid!). This enzyme's function is to cleave out the tRNA from larger pieces of RNA sequence. Not surprisingly, to me, the proteins are inert by themselves and it appears that the RNA does all the work. To prove this fact researchers put the stripped down RNA in a high-salt solution (like in a prebiotic ocean) and it still cleaved the tRNA (Guerrier-Takada et al., 1983). Another instance where RNA the length of a viroid performs a central role in the formation of tRNA without the aid of specialized proteins. Here we have more concrete evidence that my conjecture about viroids being the source of tRNAs and of important enzymes at the beginning of Life is true.

22 THE RISE OF RIBOSOMES

Viroids these random chains of RNA would replicate exponentially whenever the emergent recursive cycles would take them to the proper pH level. Lacking any correction mechanisms they would also mutate exponentially. Of course we would expect replacement mutations, deletions, and duplication mutations of its individual bases, groups of bases, or even up to its entire "genome." The later one is the quintessential most important for the next stage of Emergent Chemical Evolution.

Emergent Chemical Evolution (ECE) would have produced an innumerable amount of polypeptides that would have eventually united with the innumerable amount of, mutated by duplication, viroids. Back in 1969 Masayasu Nomura observed that ribosomes spontaneously assembled from their constituent parts (Nomura 1969). That means that without any other molecules to put them together or provide them with energy, to do so, they formed a ribosome. Many experiments have been performed that show ribosomes still work even with a majority of its proteins missing (Stryer 1988). Here are two other instances of modern, published, and corroborating evidence in favor of ECE. So according to this evidence, when the right mutated viroids formed they would have spontaneously become the first ribosomes.

Once I had seen the resemblance of viroids to tRNAs I looked at the entire shape of the viroid. When I saw the stretches of base paires with the periodic circular areas it immediately reminded me of the layout of a ribosome's RNA secondary structure. Recognizing and conjecturing that both tRNAs and ribosomes emerged from viroids makes perfect sense. Both originating from viroids means that they would automatically have had complimentary structures making it easier for them to mate. Once the viroids stumble onto these complimentary structures they automatically have the information to make tRNAs and ribosomes coded in their RNA. Viroids are non-coding RNA that have the unique ability of replicating on their own; who also share the conformational shapes of tRNAs and ribosomes; something not seen in any other "living" organism.

What is seen, in every living organism today, in the genes that form rRNA, mRNA and tRNA are stretches of base sequences within their introns that match viroid base sequences (Dinter-Gottlieb 1986). Not only are viroid sequences between the components but, surprisingly, there are always one or two tRNAs sequences between the components of the ribosomes. This is telling about how tRNAs evolved and where they came from but it is not the key take away lesson. The key take away lesson is that all these four to five components are excised from a single RNA strand that contains all of them together (Stryer 1988). This is exactly what we would expect to find if viroids were the original replicating molecule. Again concrete physical evidence that was peer reviewed, published, and written by experts that supports ECE.

As I mentioned earlier in this book, when arguing against a self-replicating RNA molecule as the origin of life: we should still find self-replicating RNA if it was able to self-replicate in such harsh conditions; especially when the conditions have steadily improved. I had also argued that we should find evidence of such a molecule in the replication systems of todays organisms. Instead, today, we find evidence for both of these ideas being exactly true only for viroids.

Now these ribosomal subunits gave ECE a big boost.

These first ribosomal subunits would have united amino acids indiscriminately; because the RNA strands would be void of any information to make useful proteins yet. And the ribosomal subunits would also lack the control portions of today's ribosomes. Those parts would evolve later, but with their absence would cause the ribosomal subunits to act as protein-assembling-enzymes. Many experiments have been performed that show ribosomes still work even with a majority of its proteins missing (Stryer 1988). And, because they acted as enzymes, they made ECE produce random sequences of proteins at a much faster rate than initially.

This new protein assembling enzyme will exponentially produce novel proteins that will introduce new pathways for ECE to follow, leading eventually to full encoding systems. When a 50s like ribosomal subunit evolved to attach itself to the 30s like original ribosomal subunit a complete ribosome was produced. Then this could interpret a sequence of RNA and also transcribe it into a protein.

Again I went looking for evidence of my conjecture and found it in an article describing how 50s subunits by themselves do not self-assemble, but with the 30s subunit they do (Nashimoto and Nomura 1970). Meaning that the 30s subunit came about first and then the 50s subunit evolved later attaching itself to the 30s subunit. ECE continuously produces random proteins that are not enzymes on their own; but proteins they are, one large molecule made up of many different amino acids. This means that these proteins had many points at which to attach themselves to other proteins, RNA, or even ribosomal subunits.

ECE produces trillions upon trillions of proteins that in turn either made other molecules-like sugars, fatty acids, and other molecules- or they attach themselves to other molecules. These proteins randomly attach themselves to every molecule they came in contact with. If a protein attaches itself with enough amino acids- by hydrogen bonding- then it is almost impossible to separate these two molecules by any physical means.

If the protein only attaches itself by just a few amino acids, these will not be enough to overcome the physical forces that separate the molecules. And so the molecules separate. Now the freed protein can attach itself to another molecule and then see if enough connections can be made with that molecule. And this process will keep repeating itself over and over again until a match is made or the protein is recycled.

Once one of these proteins attaches itself to a protein, or other molecule that fits with it perfectly, it would be very hard to separate them. So when a protein that fits the 30s like subunit and attaches itself and adds functionality to it, it will stay attached; therefore taking it one step closer to being a full ribosome. Modern ribosomes have at least 10 whole proteins and several split proteins associated with each of its subunits, 30s having the most. This is another reason that ECE is a much simpler and deterministic way of ribosomes emerging. Then life, through accumulation of addition mutations, would step by step gain the information necessary to produce these proteins in its genes. This we will discuss further later on in the book. But at this point, you have all these ribosomes that are laying around just waiting for another molecule to attach itself.

Now Transfer RNA (tRNA) is a relatively short strand of RNA that has naturally folded onto itself in a way that makes it fit onto a ribosome. It also has a short stretch of RNA that sticks out of it. This short stretch of RNA on different tRNAs will have different sequences of nucleic acids. These different sequences of nucleic acids will cause the tRNAs to attach themselves to specific amino acids. When the tRNA attaches itself to an amino acid that matches its sequence it then drags that amino acid with it to the nearest ribosome. The tRNAs' folded conformation makes it fit into the ribosome causing it to take a conformation that unites the amino acid to the already forming string of amino acids that it is attached to.

Now I have just described how an RNA- to-protein system developed in the prebiotic oceans. This system- one of many- was produced by ECE which produced innumerable random sequences of amino acids- polypeptides. With the presence of emergent recursive currents these long molecules were churned until they came together. Once together they were very hard to separate by mere physical means.

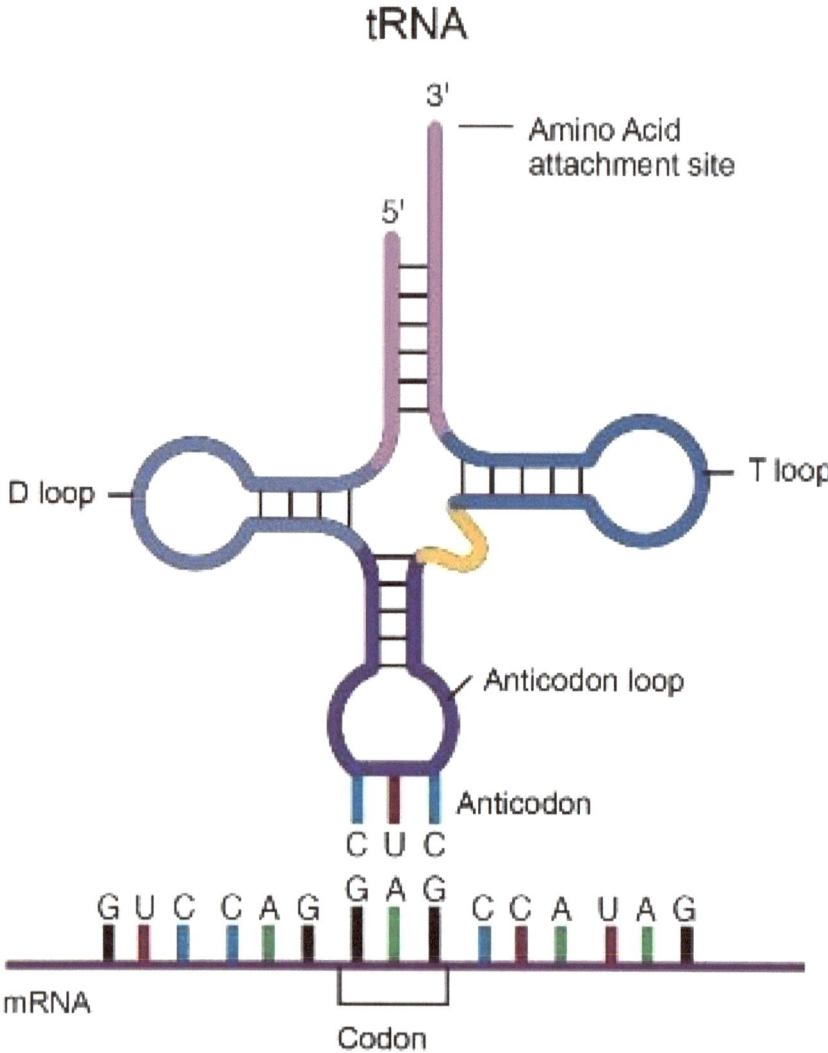

Figure 13. Transfer RNAs are short strands of RNA that spontaneously take this shape after being cleaved by an mostly RNA enzyme the same length as a viroid.

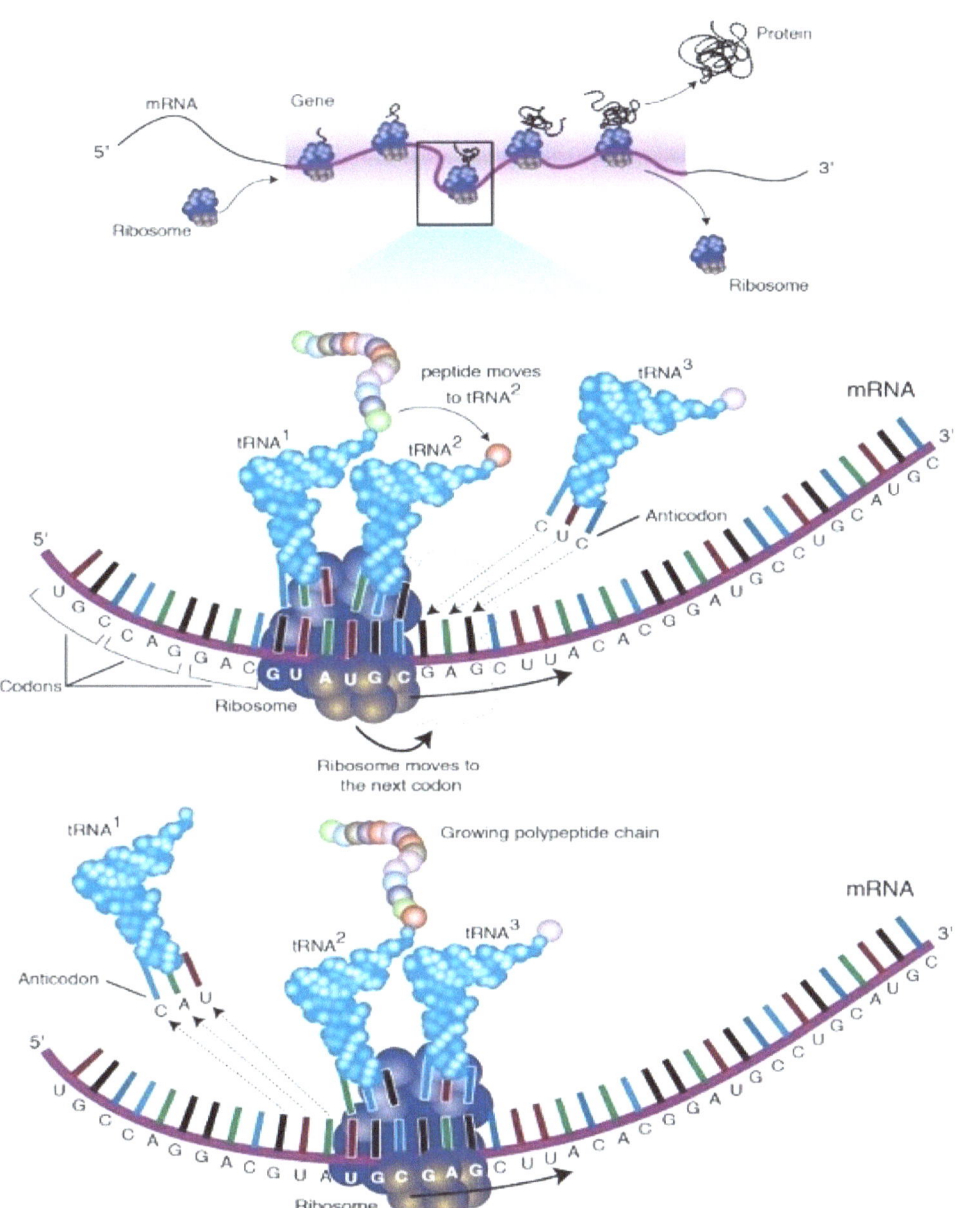

Figure 14. Ribosomes would have attached themselves to whatever random sequence of RNA it came across and starts producing proteins based on this RNA strand's sequence.

23 I'M RIGHT HANDED, WHY ARE MY AMINOS LEFT?

Now, within these emergent recursive cycles (ERCs), there would have been trillions upon trillions of polypeptides. And these polypeptides represent trillions upon trillions of possible enzymes that would have catalyzed different kinds of chemical reactions. And at this point, we must realize that the number of different combinations possible- using the twenty amino acids that life uses today- is boggling. But according to analysis conducted on meteorites that are older than the solar system itself, over all they contained over seventy different types of amino acids; thus, Emergent Chemical Evolution (ECE) had an even greater number of amino acids to work with.

    Not only did it have a greater number of different amino acid types to work with. It had an even bigger number, if you also look at the handedness of each amino acid as well. There are two ways to arrange the atoms of an amino acid. One way is called the right handed arrangement and the other the left handed. Now, life on earth only uses left handed amino acids, but this does not mean that ECE had to limit itself to left handed amino acids. The Trisapient Homolog Process (THP) would have very easily incorporated any right-handed amino acids. What matters to THP is the final arrangement of the reactive site. It could have produced very exotic combined left and right handed amino acid enzymes; therefore, the possible number of different enzymes that would have been produced by ECE is also mind boggling.

Now, the problem is that today all of life's amino acids are left handed. This is simply due- I believe- to the THP only being able to produce reactive protein sites capable of only synthesizing left-handed amino acids. So if man comes and replaces a right handed amino acid in a protein thereby changing the final conformation of the active site that protein (enzyme) would become inert. Not because right-handed amino acids cannot make active enzymes but instead because they have a different orientation. Life just stumbled onto a way of making left handed amino acids first, and, as always is the case, the life form that solves a problem first has the advantage and thus survives better; or It could have been that energetically speaking the left handed solution was the easier solution to arrive at first; or that the sequence for coding a left-handed-amino-acid-synthesizing protein into the RNA was arrived at first (my first choice) and there was no evolutionary pressure to arrive at a right handed solution after that.

But, it is just as likely that right handed amino acids cannot form viable proteins and thus cannot give rise to life. So if this is the case, and if one of the trillions upon trillions of polypeptides attaches itself to a right handed amino acid. Then that polypeptide would fall off by the wayside inactive, like an extinct species, and would eventually be recycled. Thus, if any one of these reasons is true, life would have only evolved ways of producing left handed amino acids.

What is most important to note here, is that all of these reasons- for amino acids being left handed- are compatible with ECE and are totally materialistic. Looking at ECE from this materialistic and evolutionary perspective, in light of THP, it is easy to see how easily and quickly important enzymes would have emerged with left-handed amino acids.

For an excellent example of how easily and quickly important enzymes can form, let us look at the enzymes called proteinase. Today some proteinase enzymes break down proteins in the lysosome. Some of these proteinase enzymes are polypeptides that are built from only 128 amino acids. That such a small number of amino acids are all it takes to make such an important enzyme perfectly illustrates how quickly and easily ECE, through THP, would have made them. And, what is even more amazing is that these small enzymes have the unique ability of working only within a solution of a specific pH. The same pH as that contained within the lysosomes; a very acidic 5. I posit that within the ERCs of ECE, these proteinase like enzymes would have hydrolyzed any compatible protein within their vicinity; of course, only after arriving at a layer of ocean with that enzyme's activating pH.

But how much time would it have taken a recursive system like ECE to evolve such a polypeptide with that length and catalytic ability? I can assure you- with the amount of molecules that would have been floating and diving within that system- not very long indeed. Within such a system it would not have taken very long, at all, for any type of enzyme to evolve for that matter.

Proteinases are highly specific enzymes and are sorted into six different groups depending on- among other things- how and what type of bonds they break. Proteinases are also grouped by which two amino acids make up the bond that that enzyme can hydrolyze. Proteinases are also grouped by which pH level will cause them to hydrolyze a bond. Proteinases are also grouped by the number of amino acids that compose them. And proteinases are also grouped by- what is most important of all- which co-factor they require to catalyze a reaction.

The fact that many enzymes require co-factors to work is another point in favor of my theory and another one against theirs. Co-factors are those things that an enzyme needs in order for it to perform its catalytic reaction. Co-factors can either be other proteins, different chemicals, or a variety of metal ions. The reason this favors my theory is because only in an aqueous environment, such as the ocean, could the co-factors be so readily available for those enzymes to use. And the oceans are the only environment able to supply enough metal ions-especially iron ions of the proper charge-needed to carry out just one catalytic reaction; let alone, for all of the millions of catalytic reactions that were necessary to kick start life.

The reason I used proteinase as an example was to demonstrate that it does not take many amino acids to make a biologically important protein. And how easily ECE would have setup a selective recycling system. The specificity of these enzymes has a secondary benefit of not allowing these enzymes to indiscriminately destroy all proteins that are produced by ECE. With a selective recycling system in place ECE was able to go on indefinitely. This would allow ECE to eventually come up with an RNA generating cycle that would have taken advantage of the oceans pH gradients. And the other thing about proteinase's specificity is that they do the exact same thing in living systems today. We do not have to speculate if proteins like that can work mixed in with other enzymes and biologically active chemicals in an aqueous environment; which is another aspect of ECE that can be observed today.

24 THE INDESTRUCTIBLE STAGE OF LIFE

I think that the way we discovered viruses may forever color our perception of them. They are so small- that if they had not caused disease- we might not have discovered them yet. Their size is indicative of their nature. They are the part of life that inhabits the molecular spaces left between non-living chemicals and cellular life. This is not only their niche but it is also indicative of their origin.

Once Emergent Chemical Evolution (ECE) had produced viruses, it had created perfect mutating agents. We now know that the reason that it is so hard to make a vaccine for the common cold is because it mutates so quickly. What this means is that this virus has so many replication errors during its replication that it evolves almost immediately. Evolves enough that the next time the same individual encounters this virus his immune system does not recognize it; so that person becomes sick again.

This lightning speed of evolution is not seen in any other area of life. This is of course due to the virus's very fast rate of replication. Which also is at a rate not seen in any other area of life. In turn this rate of replication is of course due to the miniscule size of viruses. Which is also a size not found in any other area of life. In fact this size places viruses closer to the size of molecules than it does to that of cells.

The molecular size and simple construction of viroids convinces me that ECE inevitably would have produced a viroid instead of a self-replicating RNA molecule. A self-replicating RNA molecule would not only be extremely complex and bulky there is no evidence that such a molecule has ever existed. Not even a remnant of it exists in any of the known replication systems in any area of life. Yet the self-assembly of the tobacco-mosaic virus was first demonstrated way back in 1955 by Heinz Fraenkel-Conrat and Robley C. Williams of the University of California at Berkeley. But it was not until they had the necessary techniques and equipment that in 1978 P. Jonathan, G. Butler, and Aaron Klug published an elucidation of the mechanics of the process, and in this article, elucidating the process, they stated that the most important factor in the self-assembly of the virus is the pH of the surrounding medium (Jonathan and others 1978). This article I consider another confirmation of the viability of ECE being the simplest and most probable path to the first cell.

The ability of viruses to self-assemble from their constituent parts is a very handy ability that was necessary for life to continue at the time of the heavy bombardment. Those hurtling objects were incessantly impacting the oceans churning up the sediments, the water, and any unfortunate viruses. When they impacted the sediments they would release any proteins, amino acids, or viruses that settled at the bottom of the ocean and became trapped. These impacts would ensure that the proteins, amino acids, and viruses that might have been lost for, possibly, eons to life were still available. This was a highly valuable action of the heavy bombardment that without it life would have taken much, much longer to start.

Just as great as the heavy bombardment's role in uniting amino acids was, its role in keeping in flux the components of Emergent Chemical Evolution was greater. Certainly many of the early viruses where disassembled by the impacts, but this was not a problem for them. These life forms which were closer to being inanimate matter than the fragile, and what we currently call, "living" cells. The viruses would have simply reassembled as soon as they found themselves in the proper pH level thanks to the unstoppable emergent recursive cycles (ERCs) of the oceans. Their components would have spontaneously taken the correct conformation and the intermolecular forces would have caused them to unite again. This almost nightmarish ability was what ensured that life did not blink out every time there was a major impact. The size of those objects ranged from small, several feet wide, meteorites to- several hundred miles wide- comets and asteroids. Think of the devastation one of the later objects would have had on the entire planet upon impact.

None of those supposed protocells would have been able to withstand them; just like that life would have had to wait a very long time for another improbable event to occur in order to form another protocell again. The zombie like viruses would have just picked up the pieces like nothing had ever happened. To those molecular zombies there would not have been much difference between the impact of a large object like an asteroid and that of a stirrer in a beaker; the effect and damage would have been comparable.

This scenario completely matches the observed fossil record. The evidence for the heavy bombardment is undeniable but not a problem for the emergence of life as some had believed; as it turned out it might have been the most crucial event for life. 600 million years of stirring the primordial pot. Restarting any false starts or stagnations. Breaking free any RNA that got irrevocably trapped in their protein coats.

Emergent Chemical Evolution would have made it possible for various types of viruses to self-assemble. ECE would not just simply stop working once a viable virus emerged. It would produce so many types of proteins and RNA segments that it would be highly improbable for ECE to have just made one type of virus. ECE would have just as easily produced helical shaped viruses as polyhedral viruses, but we have already established that the viroid was the very first.

Once ECE has produced the first virus we have an agent that not only replicates at an astounding rate but one that also mutates at an astounding rate. These mutations will of course produce not only unique viruses but unique proteins. You see, so long as the virus creates proteins that encapsulate its RNA it will be able to conserve itself as well as the new information created by the mutations. These mutations would consist of simple addition, deletions, or repeats of base pairs or even whole genes. Now- considering the rate of replication and mutation of viruses- it would not take long, in light of THP, for enough mutations to accumulate to create new and novel proteins.

Now these new and novel proteins are the key for the further evolution of not only viruses, but life itself. Of course there is the simple consequence of a protein eventually evolving enough to create a totally new and different protective coat thereby creating a new virus. But what is most important for the evolution of viruses- and as a consequence for life itself- is when these novel proteins evolve enough to mimic proteins that are involved in the protein translation process. These later proteins would provide a survival advantage for the viruses that encode for them first. It would make the replication of these viruses just a little more favorable than those that do not.

This process not only helps explain how new genetic information is created. It also explains how in the genetic code some proteins- that are needed for the translation of proteins- were encoded long before other parts were. It took longer for the right sequence of mutations to accumulate in order for the other translation protein's information to be added onto the RNA sequence. This is also a much simpler explanation for what we see. Much better than positing that life had an impaired protein encoding and translation system working flawlessly until a better one evolved.

In their musings over the origin of viruses, geneticist claim that viruses must have originated before the evolution of full fledged single celled organisms. Why do they say this? It is because viruses come with many kinds of DNA and RNA replication systems. Their genetic analysis of a broad range of viruses show that a majority of them share genes that are central to their replication and structure; that are completely missing from cellular genomes. Because they believe that viruses originated from cells, and since there are viruses that only use single strand RNA systems for their replication. They believe viruses must have created the DNA system and then introduced it into the cellular lineage. They believe that all of these viruses then maintained their genetic purity from the precellular stage of life before there was extensive genetic mixing (Koonin and others 2006). I believe this to be a misinterpretation of the data.

If viruses originated from cellular genomes, then they should not be missing any genes that are central to the replication and structure of cells. Especially when these are the genes, like in the majority of viruses and cells, are the ones that are most preserved. This action not only preserves the identity of the organism but also allows it to maintain its ability to replicate and quite simply survive. Besides, in all other areas of life, the evidence to date shows that once life has a solution to a problem it stays with that solution or builds upon it by mutation. I have yet to come across any evidence that shows life ever dropping one solution to a problem and de novo adopting a totally different solution.

On the other hand, if there was an evolutionary paradigm that makes life switch from one perfectly good encoding and translation system to another "better" one. Then, are not the mitochondria a couple of billion years late in switching? Especially when they have the opportunity every time their host cell divides and reproduces itself to adopt the host's encoding and translation system. But instead, what we see is that the mitochondria preserve themselves. How can we be sure of this?

The mitochondria not only have a different genetic code, they also have their own ribosomes, and transfer RNA (see figure 15); neither of which are shared with the host cell (viroids are a better explanation for this). This clearly demonstrates that early in the evolution of cellular life there was no genetic mixing. The ability of genetic mixing we see today between single celled life forms clearly had to evolve- much like sex- much later in the history of life.

The mitochondria also demonstrate that- central to

the preservation of functionality and heredity- there has to be an unbroken chain of genes that reaches far back to the beginning of Life. And nowhere do we find any such break in the sequence of genes in the chromosomes of eukaryotic cells.

The ability of viruses to self-assemble from their constituent parts is a very handy ability that was necessary for life to continue at the time of the heavy bombardment. Those hurtling objects were incessantly impacting the oceans churning up the sediments, the water, and any unfortunate viruses. When they impacted the sediments they would release any proteins, amino acids, or viruses that settled at the bottom of the ocean and became trapped. These impacts would ensure that the proteins, amino acids, and viruses that might have been lost for, possibly, eons to life were still available. This was a highly valuable action of the heavy bombardment that without it life would have taken much, much longer to start.

Just as great as the heavy bombardment's role in uniting amino acids, was its role in keeping in flux the components of Emergent Chemical Evolution. Certainly many of the early viruses where disassembled by the impacts, but this was not a problem for them. These life forms which were closer to being inanimate matter than the fragile, what we currently call, "living" cells. The viruses would have simply reassembled as soon as they found themselves in the proper pH level thanks to the unstoppable emergent recursive cycles (ERCs) of the oceans. Their components would have spontaneously taken the correct conformation and the intermolecular forces would have caused them to unite again. This almost nightmarish ability was what ensured that life did not blink out every time there was a major impact. The size of those objects ranged from small, several feet wide, meteorites to- several hundred miles wide- comets and asteroids. Think of the devastation one of the later objects would have had on the entire planet upon impact. None of those supposed protocells would have been able to withstand them; just like that life would have had to wait a very long time for another improbable event to occur in order to form another protocell again. The zombie like viruses would have just picked up the pieces like nothing had ever happened. To those molecular zombies there would not have been much difference between the impact of a large object like an asteroid and that of a stirrer in a beaker; the effect and damage would have been comparable.

This scenario completely matches the observed fossil record. The evidence for the heavy bombardment is undeniable but not a problem for the emergence of life as some had believed; as it turns out it might have been the most crucial event for life. 600 million years of stirring the primordial pot. Restarting any false starts or stagnations. Breaking free any RNA that got irrevocably trapped in protein coats.

In another article they were trying to figure out how DNA and DNA replication proteins evolved. This article also points out that there are different viruses with different genetic encoding systems. Some viruses have single stranded RNA, double stranded RNA, single stranded DNA, double stranded DNA, and modified DNA. The article also explains that there are different protein families that manipulate the genetic systems of each kingdom of life which have nothing in common with each other (easily explained by THP). Being this another sign that there was no extensive genetic mixing early in the history of life. They also suggest that those systems had independent origins and that they were viral in origin and introduction. That viruses at different times infected cells with these new genetic systems and replaced the original RNA system (Forterre and others 2002).

Although the authors of the two previous articles- I believe- drew incorrect conclusions, their articles never the less provide us with modern, corroborating, and published evidence of my theory's correctness. They provide us with the evidence of viruses having different genetic systems not only from each other but from eukaryotic cells. They also provide the evidence that those viruses got these genetic systems first long before modern cells did and that they each evolved different types of- totally unrelated- proteins to manipulate the RNA-to-Protein or DNA-to-protein systems within the different domains of life. All of which my theory of

Emergent Chemical Evolution through the Trisapient Homolog Process predicts. All of this provides even more evidence that ECE better explains the origin of viruses better than the current theories do, and better fits the observed data.

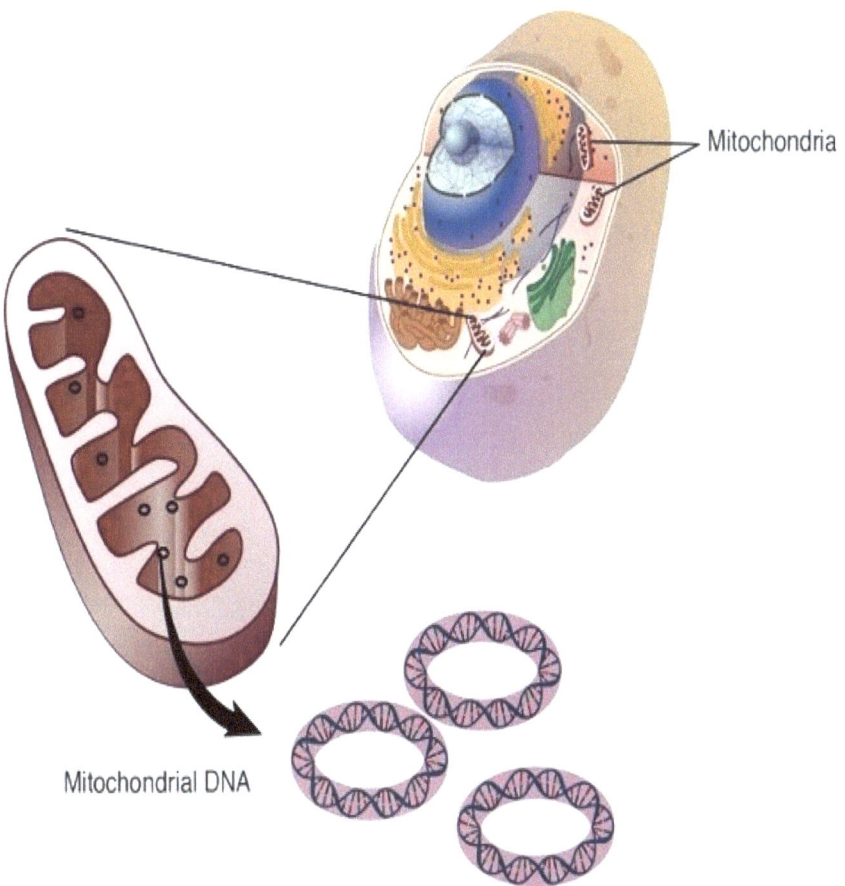

Figure 15. The mitochondrion preserves its DNA translation system and produces proteins from its own ribosomes and other proteins with the help of the host.

25 MULTIPLE STRAINS, MULTIPLE LINEAGES

We humans need to have everything in order and would like everything to fall into an easily defined category. We want the origin of life to be simple. To be one random event that took place at one specific location. And that Life- as we know it- started with only one original life form that led to the current tree of Life. And I would have loved to be the one to deliver that unifying theory, but I am afraid that that theory just does not exist. Even worse than that, I believe we have to cut off some roots from the Tree of Life and construct separate bushes from them. But I do provide one emergent mechanism that leads to the emergence of life's tree and bushes.

It would be very unlikely that Emergent Chemical Evolution (ECE) stopped working once the first viroid was formed. And it would also be very unlikely that it would have produced only one type of virus after that. Just as easily ECE would have produced a helical virus as it would an icosahedral virus or vice versa. These viruses are so simple that it seems easier to see them as products of ECE than- the yet unobserved- spontaneous breaking away of RNA or-even less probable- DNA from a catastrophically damaged cell.

Evolutionarily speaking- it is much easier for a protein to evolve into encapsulating RNA than spontaneously doing so. An RNA strand swimming in the conditions favorable for ECE would not perish. And the RNA strand would have an almost unlimited number of chances to mutate-by self-cleaving and recombining- and eventually producing a protein that would encapsulate it. With ECE there would not be just one RNA strand- happening by chance- but billions of independent RNA strands evolving at the same time thereby increasing the chances of one of them accomplishing encapsulation.

Whereas with de-evolution it would only get one chance for the membrane to encapsulate a compatible RNA segment. If it did not succeed, if we keep with all the erroneous assumptions, the RNA segment would perish. I think that the probability of the same type of cell suffering the same catastrophic event and that the same segment of RNA breaking off again-after a failed attempt- is way lower than if it had occurred by ECE. Not to mention- what is the probability of that same portion of membrane being both compatible with the RNA strand and another cell membrane? Or that that membrane will not only bond with it but also give it the right conformation to release the RNA into said cell in the first place?

What mechanisms would cause a portion of a membrane to spontaneously encapsulate a segment of RNA? At what point will a cell's machinery change enough to stop being a cell and produce viruses? How many generations will a cell produce after having its genetic information changed so drastically that it will eventually produce viruses? Would not all those changes to a cell's genetic information be detrimental to its ability to replicate? So detrimental, in fact, that it would cease to exist before it evolved enough to produce a virus? Doesn't the hypothesis of viruses originating from living cells create more questions than it answers? Yes.

Viruses were the first organisms to emerge from the mechanism of ECE. No other conclusion can be drawn. Especially when that conclusion is based on the fact that the simplest of viruses have the smallest genomes and that those genomes only produce one protein to encapsulate their RNA. And also these proteins are different from- and unrelated to- the proteins encapsulating other types of simple viruses. Whereas, one would expect the viruses to converge on the same protein coat if they de-evolved from the same cell type that gave rise to all complex life. And what we would also expect to see are viruses using the same, or related, proteins to manipulate their genetic information as the proteins seen, doing the same job, in the rest of the kingdoms of life. But not only are these proteins unrelated to the rest of life (again easily explained by THP). They are also unrelated to the same type of proteins found in other types of viruses as well (Iyer and others 2001).

Now this last point, for many, will be a point of contention. Because some of you will say that the different viruses could have originated from extinct cell lineages or at very different points in the evolution of eukaryotic cells. Well both of these arguments still suffer from the same problem as before. They would still have genetic information that could be linked back to the rest of life. Either of these originators would still be descendants of the original cell that gave rise to the rest of life. And it would also have related proteins no matter how much they would have diverged. And that those proteins could not have diverged so much- because had they- at some point they would cease to perform their original function. And then they would have become unfavorable mutations and thus lead to those viruses' extinction.

The more and more you look at the origin of viruses from already formed cells the more untenable it seems. We have clearly demonstrated that ECE is a much simpler and evolutionarily congruent means for the abiogenesis of viroids. And we have also demonstrated that ECE would have produced many types of viruses. What we need to do now is explain why there must be bushes alongside the tree of life.

Well in order for me to explain that to you, I have to explain another facet of my theory that until this point has only been implied. That viruses evolved to form cellular life. In the next couple of paragraphs I'm going to explain to you just how this happened.

According to my theory of ECE, life was initiated once a viroid was formed. Now here is where it gets a little tricky. There are several types of capsid (protein only coat) viruses that could have formed from ECE. Each type of virus that is formed can be classified as a start to life. They each can because each virus is going to evolve. Evolve either into a type of cell (bacteria, archaea, protozoa, fungi, algae, plant, or animal) or be the first ancestor of any of the modern viruses. And since I have already established that viruses are both alive and dead, they are- none the less- still part of life.

Since the most popular view on the origin of viruses is that they de-evolved; from living cells. The current tree of life has its roots come from the origin of archaea and protozoa. Viruses are just a weird branch on the tree of life. I think we have to redraw the tree of life. Drawing on my theory posits that different and unique viruses emerged from ECE; some of these viruses evolved into more complex viruses, and then merge together to form the ancestors of modern cells. I also deduce that due to their unstable genetic storage system- i.e. single stranded RNA viruses. And because of that instability some viruses did not evolve any further than being viruses. And, as long as their environmental niche still exists, they will survive keeping their unstable genetic storage systems.

These viruses' genetic apparatus are so unstable that their genomes cannot grow any further. After a certain point they become nonviable and cease to produce any viruses. That is why it appears that some of these viruses have maintained their genetic purity from the beginning of life. Because they are unable to evolve any further. And that those viruses that evolved a genetic system a little more stable than that, evolved into archaea, protozoa, fungi, and algae. But that their genetic systems were not stable enough for these "modern" cells to evolve any further either.

The Trisapient Homologue Process would have allowed many versions of the same reactive site to form inside very different amino acid sequences. What matters, in the figurative sense, most to the Trisapient Homologue Process is the arrangement in 3D space of the amino acids that form the reactive site. The other amino acids, that help form the rest of the enzyme, need to take the necessary conformation for the reactive amino acids to be in their proper place in 3D space. Now it could be that the arrangement for a particular enzyme was to have two alpha-helices with one beta-sheet in between them. In this particular arrangement the enzyme needs to find itself in a pH of 5 in order to get the right amino acids in their reactive positions.

On the other hand, let us say that another enzyme- with exactly the same reactive site- was formed by none-reactive amino acids arranged with two opposite turning alpha-helices and one beta-sheet in between them. The difference with this enzyme is that in order for the reactive site to take its rightful conformation the enzyme needs to find itself in a different pH of 3. THP tells us that there are hundreds of millions of different ways for that reactive site to arrange itself; or for life to stumble onto that sequence of amino acids. This also means that this same reactive site can be made to work under very different pH environments. This would be due to THP very easily recreating the reactive site but the rest of the protein would be composed of a different sequence. That very different sequence would more than likely require a different pH in order to take the conformation that would make that site reactive. Another one of life's mysteries solved by Emergent Chemical Evolution.

What this also implies is that when a virus, through THP, gets enzymes that works at a certain pH. That virus acquires the ability to live in that environment. The virus's ability to multiply exponentially and acquire mutations so quickly would make it develop, much quicker, its progeny to inhabit; ever more different environments. Those progeny that developed enzymes that worked in disparate environments would become extinct and be recycled. Eventually organisms that would have all of their enzymes optimized for one type of pH would eventually evolve into todays extremophiles. Then as the atmosphere became more oxygenated those extreme environments would recede and only those organisms fit to live in them, and lucky enough to still find themselves there, would survive. Here I have easily explained the, until now very perplexing, ability of viruses to function in environments with very different pH levels had evolved.

The other implication of different viruses giving rise to distinct cell linages means that there is no Last Universal Cellular Ancestor (LUCA); but a LUVA (Last Universal Viroid Ancestor). Which is the viroid and this is because all viruses evolved from copies of the original viroid. The viroid cuts off all the other roots from the Tree of Life and brings the tree instead to a very sharp point.

26 A PRECELLULAR ECOSYSTEM IS FORMED

Viruses are the quintessential mutation machines. But what is it that mutates? Obviously what changes is the genetic information coding for proteins. And how is this beneficial to a virus? Simple, it changes its essential proteins in novel ways that can make them work better. And when it comes to mutations involving duplication and addition mutations, new proteins can be created that can give a virus an advantage over all the other viruses. These new proteins can add new functionality to virus's replication system and/or add beneficial structures to the virus itself.

New replication functionality has obvious benefits that play a role in the further evolution of viruses toward being cells. And there were other replication mutations that are not so obvious that aided viruses towards that goal, too. Both of these types of mutations will be discussed later. Right now the implications of beneficial structures being added to viruses needs to be discussed first. Not only for its importance but because this type of mutation occurs more quickly and more often.

The incredible rate at which viruses replicate is the same reason that viruses mutate so quickly. These mutations are, of course, the changing of the genetic information that codes for the proteins produced during the replication of the virus. Now these mutations can lead to additional proteins being produced. These new proteins could be beneficial, detrimental, or neutral. Well the detrimental proteins are the ones that would in some way or another hinder the virus's replication; and would lead to, that viruses progeny to multiply very slowly, or not at all. Eventually leading to that virus's line becoming extinct.

A neutral protein is one that neither helps nor hinders the virus. But that does not mean that in the future it will not become a very important protein with further additional mutations. These neutral mutations are the most important type of mutations because they allow new proteins that are totally unrelated to the old proteins to emerge. Once these neutral proteins mutate enough to become functional proteins and are beneficial to the virus, these neutral mutations then change into beneficial mutations.

These beneficial mutations would produce totally new proteins that have totally unique properties. Properties that would make the virus perform additional activities while it replicates or after. Some of these, additional activities, would include the new proteins incorporating themselves into the membrane to make it bigger, stronger, or even give it new functionality. And some of these new proteins would give the virus the ability to internally produce, process, or breakdown other molecules after it has finished replicating. Both these mechanisms can be observed in viruses today in the Megavirus and Mimivirus (Claverie and Abergel 2010).

When a new protein embeds itself into a membrane it automatically makes it larger. The other proteins of the virus' coat assemble with the new ones like jig-saw puzzle pieces. The new proteins can make the membrane stronger by providing it with more rigidity or by protruding out and keeping things away or reaching out to attach it to others. Now these new viruses would have the space to capture new, extra strands of RNA. That would give that virus new functionality that it did not have before or was even close to evolving. This mechanism would have been a big boost to the further evolution of viruses.

The idea- that protein coats would expand with

different new proteins and capture other RNA strands- was something I came up with as a natural consequence of my theory. And when I went looking, mind you just for evidence of multiple proteins covering a single RNA virus, I found evidence for all of it. The modern examples of this mechanism are the RNA viruses: polio with a shell composed of sixty copies each of four different proteins; and, the reovirus who's coat is composed of three different kinds of proteins, but who's RNA genome is composed of ten different linear double-stranded (+/-) RNA molecules (Strayer 1988).

The acquisition of new genetic information in this way would have made it possible for viruses to change very quickly. These new genes would have produced proteins that changed the fundamental nature of the virus. Very quickly proteins that could manipulate the genetic information would have emerged. Like giving the virus the ability to convert its single stranded RNA genome to a double stranded RNA genome. Or convert its single stranded RNA genome to a single stranded DNA genome like retro viruses do. Or convert its single stranded DNA genome to a double stranded DNA genome. And that, ladies and gentlemen, is the explanation for why we find viruses with different genetic systems. And why, thanks to THP, they have unrelated proteins manipulating those different genetic systems.

Once the genomes were converted into these more stable molecules; this would have allowed larger viruses, with many more proteins, to evolve. Evidence for this mechanism was found in 2012 when an RNA-DNA hybrid virus (RDHV) was discovered. This single-stranded DNA virus contains information for coding capsid proteins only found in single-stranded RNA viruses (Diemer and Stedman 2012). Once again ECE perfectly explains what the authors of this article could not.

The RDHV article also provides me with evidence for the existence of single-stranded DNA viruses that are covered only by proteins. I did not know that single-stranded DNA viruses covered only by proteins existed. My theory of ECE led me to the conclusion that they must exist. So I went looking for evidence of it and found this article. Again, in my mind, this is another prediction that my theory made confirmed by this article and this article, being peer reviewed, validates my theory even more. Some of these new proteins would have also embedded themselves in the membrane. This would have also made it possible for particular molecules, electrons, or ions to pass through the membrane. The proteins that provided these new functionalities to the viruses would have brought those viruses very close to becoming protozoan cells; like Escherichia coli. Which are modern cells that do not have a mitochondria but whose membrane embedded proteins are very much like that of the mitochondria. Escherichia coli gets its energy through the proteins embedded in its membrane; just like the mitochondria does within a living cell. Escherichia coli lives in a very acidic environment just like that of the oceans before there was an oxygenated atmosphere. In our guts we harbor a time capsule from the beginning of life, Escherichia coli.

The new proteins on the capsid of the virus would have caused gradients between its internal and external environments. This would have allowed the virus to setup a protein system of electron transportation on its surface. Producing proteins that would internally mimic the Krebs cycle just like those of the Escherichia coli, mitochondria, or chloroplasts, making this virus an ATP producing virus; the forefather of the Escherichia coli, mitochondria, or chloroplasts.

"But how can a virus with a protein coat be the forefather of cells with a lipid bilayer membrane?" you may ask. It is quite straightforward, let me tell ya. The proteins on the surface of the virus were producing energy just like Escherichia coli does today. This energy at first would have been immediately used by the virus and the excess lost to the environment. It would not have been long before a system of storage would have evolved. And where else but on its surface could a virus hold these new proteins because it has no other internal structures, yet.

The energy storing proteins would have been located on the surface of the virus. Here, just like today's cells, they would produce phospholipids. Lipids are fat molecules that can store enough energy to make up to

106 Adenosine Triphosphate (ATP) molecules. Phospholipids are very specialized kind of lipids that have one water loving phosphate head and two water fearing fat tails. Phospholipids by their very nature will form a bi-layered membrane spontaneously in water.

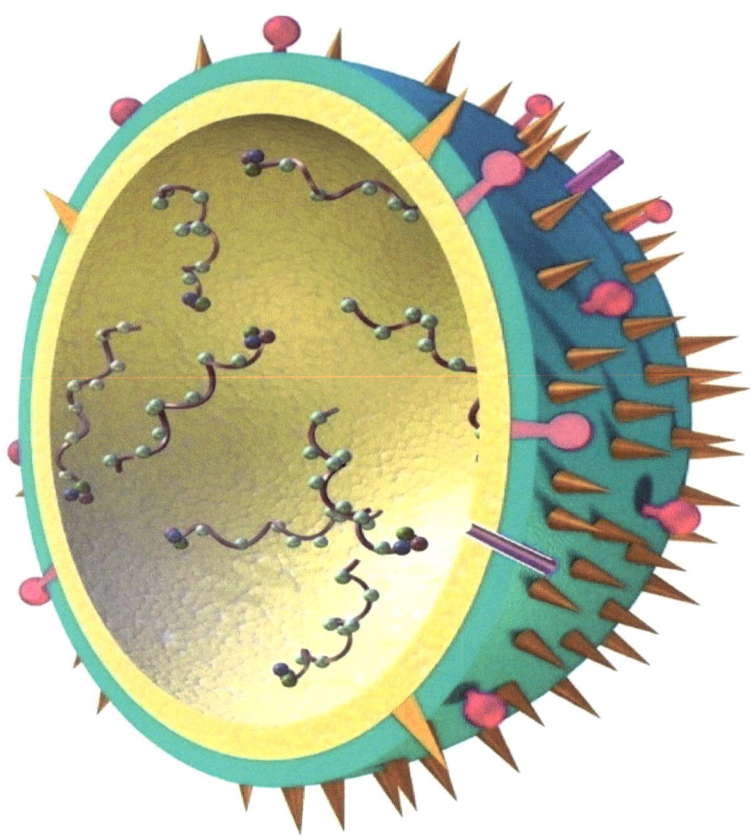

Figure 15. This computer model of the avian flu virus clearly shows all of the viral characteristics of expanding capsid to capture more genetic material, protein spikes, and protein portals.

Once a virus starts producing phospholipids, the phospholipids will move towards the virus coat and start filling in the spaces between the proteins of the coat, and embed the proteins automatically; hence, solving the self-assembly problem of the first phospholipid and protein membrane, and thus, creating the first virus with a phospholipid and protein membrane, or as it is also known a lipoprotein membrane.

When a lipoprotein membraned virus showed up it changed the very nature of membranes forever. Now the virus has a covering that is malleable and expandable in response to internal and external pressures. Now remember, all of this is not happening in a vacuum. While some viruses are obviously evolving into more complex organism, their cousins or failed siblings are still around; round to infect them, to compete with them, or even hang out until today; of course, as long as their niche is still around.

Evolutionary principles tell us that competition stems from two organisms trying to occupy the same niche in the same environment. More and more viruses were being replicated and therefore mutated. They will naturally have individual differences that will make each one prefer a different niche. Like the ATP virus, it would prefer a niche that would maximize its ability to make a gradient. Other viruses would have evolved to do different things. Like some could produce a variety of fat molecules and these would like to be in a niche close to the ATP virus and a niche that would have provided the raw materials necessary to produce those molecules as well.

The innumerable types of viruses that evolved during this period would have easily created a virus ecosystem. Each type of virus would occupy their own niche providing something that benefits the ecosystem. There would also be viruses that would infect the other viruses and thereby get a lipoprotein coat. And other viruses like in today's ecosystems would setup symbiotic relationships.

27 THE RISE OF EUKARYOTES

A symbiotic relationship is where one type of organism co-exists with another in a mutually beneficial manner. For example, a sea anemone gives shelter to a clown fish in return for the clown fish keeping it safe from butterfly fish. This particular relationship developed thanks to a beneficial mutation to the clown fish's mucus coating that makes it invisible to the stinging tentacles of the sea anemone. So you see it was a fortunate genetic mutation that made this relationship possible. It was not the active choice of either one to have this relationship and then the biology fallowed. So it must have been back then for the viral ecosystem. Evidence for this viral symbiotic relationship being possible is provided by the viral "parasite" sputnik who "attacks" Mimivirus (Claverie and Abergel 2009).

I consider the relationship between sputnik and Mimivirus as a failed symbiotic relationship. Here you have the incredible example of one virus only able to attack the other virus within an amoeba, again, only able to do it in conditions that are favorable for ECE. And the Mimivirus is so big and complex it performs functions normally done only in cells; like nucleotide and amino acid synthesis. But again the Mimivirus also needs the conditions conducive to ECE in order to do these things as well. Although this is a failed symbiotic relationship it still provides us with evidence that, under the conditions conducive to ECE, viruses will attack each other. This also gives us a mechanism for the establishment of viral symbiotic relationships.

The viral ecosystem would have been rife with symbiotic relationships between many viruses. Whole interdependent networks would form much like ecosystems do today. Except this ecosystem would have been built upon complex viruses, doing what today we consider odd jobs for viruses. Viruses would independently mutate themselves into a niche: getting distinct methods for replicating themselves (more complex vs. less complex), through THP acquiring distinct genetic coding for proteins that do the same job, and even distinct genetic triplet codes (likes that of the mitochondria).

Eventually several viruses would mutate enough to have several of their proteins evolve to: hold their double-stranded DNA better, translate DNA to RNA, self-assemble with RNA to form their own unique ribosomes, and also embedding proteins into the membrane eventually forming respiratory chains on their surfaces; hence, becoming the first cells, the ancestors to Escherichia coli and the ancestors of mitochondria.

Very quickly separate cell lineages would have evolved from the different types of viruses. These first cells would displace viruses and less evolved cells that could not compete with them for the same niche. Then these new cells would form new symbiotic relationships that would lead to different mitochondria being incorporated, or not, into the separate cell lineages. Then each cell lineage with its own original unique capsid proteins would make distinct membranes; thus, giving rise to the different types of prokaryotes and the ancestors of eukaryotes.

The difference between prokaryotes and eukaryotes comes from the viral stage. Those viruses that formed circular RNA genetic systems and did not bind them with proteins to their capsids became prokaryotes. Those viruses that did bind their multiple linear double-stranded (+/-) RNA molecules to a protein core, like the reovirus, became eukaryotes. The reovirus has ten distinct linear double-stranded (+/-) RNA strands that are attached by proteins to a core. Then when it is replicating it extrudes the copies through pores in the core (Strayer 1988); muck like eukaryotes do when their DNA is being translated into mRNA and then the mRNA leaves the nucleus through the pores of the nuclear membrane.

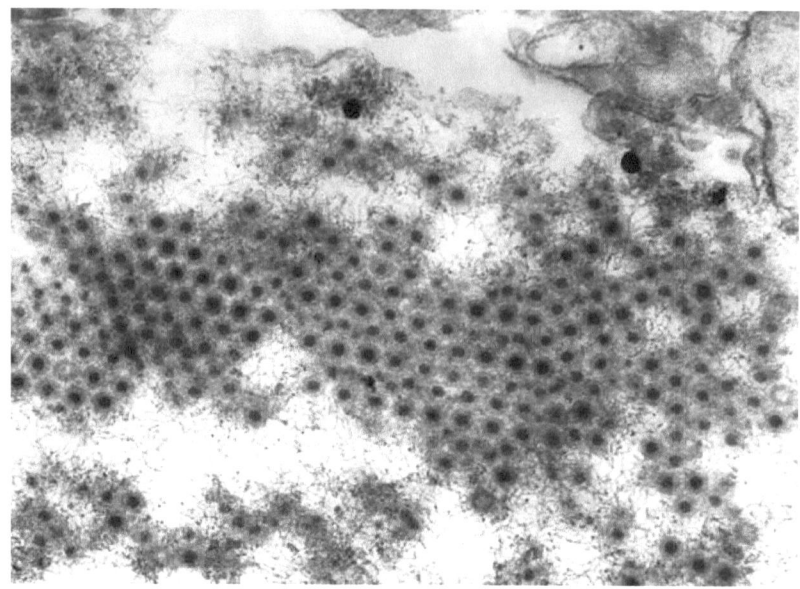

Figure 16. Here is a micrograph of Reovirus with its nucleus-like core in its center.

I posit that the ancestors of the reovirus evolved an enzyme that converted its RNA genome into a DNA genome, which was still attached to those binding proteins like modern histones. Thereby, converting ten, or less, distinct linear double-stranded (+/-) RNA molecules into the first ten, or less, chromosomes. They also retained the ability to translate the DNA into RNA and still extrude the RNA-copies (mRNA) through the pores of the core. This more stable configuration of DNA storage and translation made it possible for the proto-eukaryotes to grow a much larger and more protected genome. This larger and more protected genome meant an immediate survival advantage for the proto-eukaryotes. This included: in the meantime another layer of protection against other viruses, parallel translation of the genetic code, parallel expansion(accumulation of mutations) of the genome; and, a much larger and complex- but more stable- cell that can more easily evolve even further.

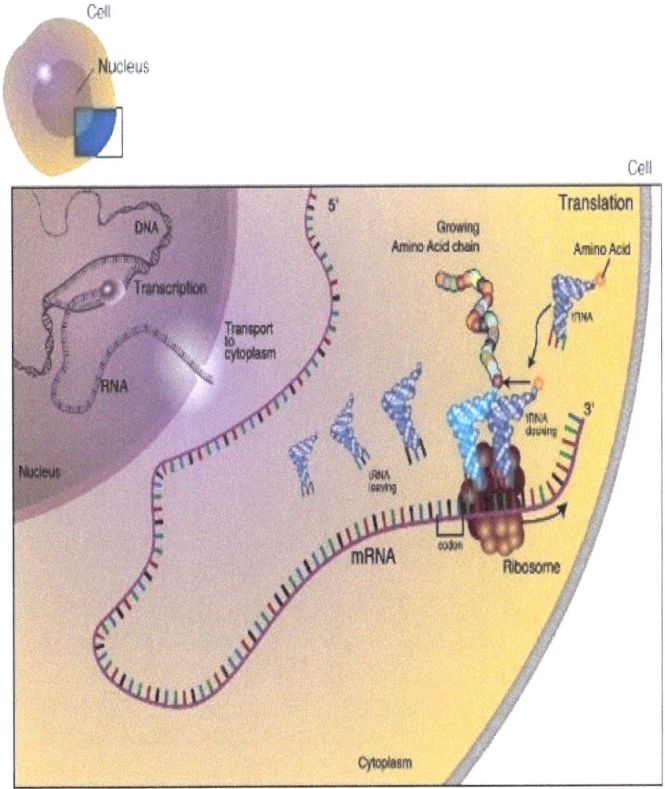

Figure 17. Here is a modern cell extruding copied-RNA from a pore of its core (nucleus). Exactly like Reovirus does when it copies its RNA genome, which it also has attached to protein anchors(chromatin) in its protein core.

Another reason I do not believe that the nucleus is not the remnants of a DNA virus invading a prokaryote is the nuclear membrane. When the nuclear membrane breaks down during replication there is a point where it is indistinguishable from the endoplasmic reticulum. Evolutionarily speaking it is very straightforward for a proto-eukaryote cell to evolve more proteins that would endow it with more internal structures over time. Such as enzymes that would convert the protein core of a virus into a nucleus and then also add structures like the endoplasmic reticulum as they do now in modern cells.

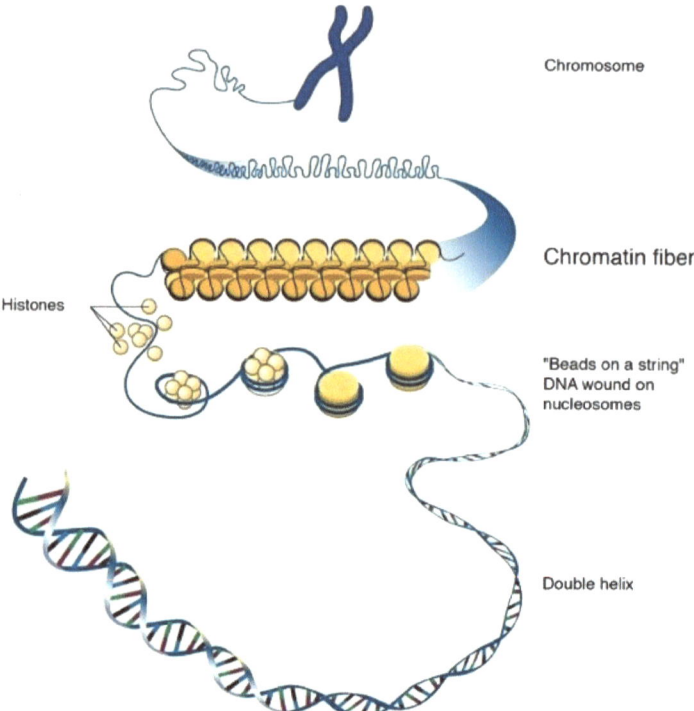

Figure 18. Modern chromosomes are much larger than reovirus's ten linear strands of RNA but both are attached to protein anchors and they both have the exact same benefits.

This conversion is especially easy starting from the reovirus whose replication mechanisms exactly matches the activities of a modern nucleus (Cullen 2003). It would not be long before this proto-eukaryote would evolve enough structures and mechanisms to then capture and have a symbiotic co-habitation with mitochondria to become an animal cell, and then later with chloroplasts to become green algae and eventually a plant cell. And this was the beginning of how life became the way it is today.

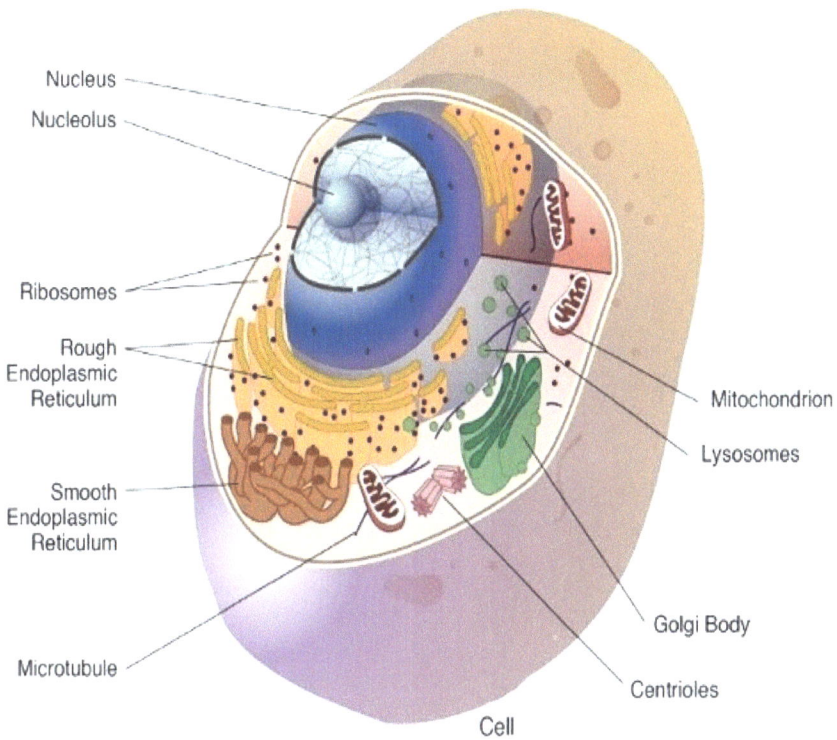

Figure 19. Looking at a modern cell's structure it is easy to imagine a Reovirus evolving more proteins that would convert its RNA to DNA and also add more complex structures causing it to evolve into a eukaryotic cell.

Going from simpler structures to more complex structures by Emergent Chemical Evolution is a totally natural and materialistic way of going from inanimate matter to viruses then to cells. The important thing to note here is that no evolutionary principles were broken or any unobserved exotic chemical reactions employed in order for viroids to evolve into viruses; then viruses to evolve into cells. Whereas going from cell to virus does break evolutionary and emergent principles, and it is even worse going from self-replicating molecule to protocell to a living cell. Emergent Chemical Evolution is a much simpler and straightforward way for inanimate matter to coalesce into Life.

ECE		RNA WORLD	
Posits	Observed Today	Posits	Observed Today
Amino Acids polymerization in water	X	RNA self-assemble spontaneously	
Ribosomes self-assemble	X	RNA self-replicates	
Meteorites source of Amino Acids	X	Enviroment on Earth produced RNA	
Viruses self-assemble	X	Protocell self-assemble	
Viruses replicate in conditions of ECE	X	Protocell replicates	
Explains Mitochondria	X	Explains Mitochondria	
Explains Chloroplasts	X	Explains Chloroplasts	
Explains different types of Genetic Systems	X	Explains different types of Genetic Systems	
Explains different viral Genetic Systems	X	Explains different viral Genetic Systems	

Figure 20. The comparison of Emergent Chemical Evolution with RNA world hypothesis shows ECE's superior explanatory power and modern evidence.

28 TYING THE EMERGENCE OF LIFE WITH THE EMERGENCE OF THE UNIVERSE

Everything we see in the Universe is here due to the emergence of gravity and other cyclic forces, so why should it be any different for Life's origin? Gravity with it's unlimited force supplied the universe with the energy to organize itself. Gravity supplied the necessary force to create the stars that first lit up the Universe. Once these stars ignited and started their fusion reactions they began producing heavier atoms. After these, and successive, stars died and released their heavier atoms into space, gravity then grouped them together to form our solar system. In the process of making our solar system the free atoms and molecules also combined to form amino acids. Once the Earth had formed, gravity kept the oceans on it, kept the atmosphere on it, and kept it spinning causing its oceans to form currents which in turn powered Emergent Chemical Evolution. Undoubtedly Gravity was the most important factor and source of energy for the origin of life.

The process of Emergent Chemical Evolution which inevitably led to the origin of viroids fills in the gap between the emergence of the Universe and the emergence of Life. As I have explained previously, the amino acids that started Emergent Chemical Evolution were produced by the same emergent forces that created the material that makes up our Solar System. It was the force of gravity that formed our Solar System and kept the atmosphere against the Earth. It was the atmosphere's carbon dioxide that imparted the pH to the oceans that let Emergent Chemical Evolution work. Gravity also powered the oceanic currents that created the Emergent Recursive Cycles that powered Emergent Chemical Evolution. So the same physical forces that make this Universe ideal for the production of black holes are also the same forces that make it ideal for the emergence of Life. And these forces are what tie the Emergence of Life with the Emergence of the Universe making Life just another one of the universe's many emergent phenomena.

Having discovered the Trisapient Homologue Process we can now figure out ways of making the enzymes in Life better; because life just kept on using whatever protein worked. Which was, usually, whatever solution it stumbled onto first. With the patented process of THP, we can now re-orient the reactive sites of proteins in order to make them better performing ones. Which could lead to making more efficient plants that could take up more carbon dioxide; doing two things at the same time: one growing more food and two taking in more carbon dioxide from the atmosphere ending at once global hunger and climate change. We could also design better medications to fight cancer, help us live longer, and live healthier lives. The possibilities are endless for this patent.

Since the Renaissance, science has provided us with the truth of reality. A vastly different, but- in my humble opinion- vastly more beautiful explanation of how the world works than that given by any religion. Science tells us that the earth and its nature are not separate from the seemingly perfect heavens. That we humans are one with that nature, and that we and that nature are tied to the rest of the Universe; because we are all made of star dust. Thanks to my theory, we can now have- what is to me- the most awe inspiring realization: that we can look up at the night sky, take in the majesty of the Universe, know that we are tied to that majesty; and that thanks to Shooting Stars we are here! What greater gift could Shooting Stars give us than that of which they have already granted; the gift of Life itself!

Review of Evidence for Emergent Chemical Evolution

No.	Emergent Chemical Evolution Claim	Published Positive Experimental Results
1	pH Layering of Oceans	Coral bleaching at low depth
2	Abundance of Amino Acids in ancient ocean	Amino Acids present in meteorites, and comets which are the source of the oceans
3	Amino Acids spontaneously bond	Chemical caps used to build man made proteins
4	Oceans full of polypeptides	High velocity comet impact experiments make polypeptides
5	Small polypeptides unite to make RNA	E.coli uses 12 small polypeptides to start the RNA synthesis by first UNITING and then transforming a single AMINO ACID into RNA
6	Trisapient Homolog Process	NRMT2 homolog of NRMT1 have 42% different sequences
7	3D-sequential-residue Relationship	NRMT2 Homologue NRMT1 active amino acids occupy the same spot in space but not in the sequence
8	Ocean currents power chemical reactions	Mitochondria produce 32 ATP by creating pH differential creating a current of ions
9	Cell membrane unnecessary	Proteins activate and deactivate by surrounding solution's pH level
10	Non-coding RNA Appeared First	Viroids are RNA without coating, penetrate nucleuses, and do not code for proteins
11	Viroids gave rise to tRNA	Viroid size RNA molecules excise tRNA from larger RNA molecules today
12	Replicated Viroids gave rise to ribosomes	Ribosomal introns match Viroid sequences and rRNA look like multiple Viroids strung together
13	Viruses formed before cells	Viruses Self-assemble in petri-dishes; cells do not
14	Viruses evolved new genetic systems, evolved into cells, and have RNA only viruses today	Viruses have unique genetic systems different from cells and there are no RNA only cell

ABOUT THE AUTHOR

Eduardo Trisapient Hernandez lives in Miami, Florida with his wife and three children. Since childhood has been fascinated with science. At the age of twelve he began studying electronics as a hobby. He spent 20 years repairing electronics for the Army, Florida National Guard, and the avionics subdivision of the company that built the space shuttle. While working repairing electronics he earned a chemistry degree. When his avionics division was moved to Mexico and he was laid-off he, once again, returned to school to pursue a medical career, his third career. Being an atheist, knowing chemistry, and having so many years of professional experience applying scientific knowledge to do his work put him in the unique position, once he started studying how to keep people alive, of solving the Abiogenesis Puzzle.

REFERENCES

C. M. O'D. Alexander, R. Bowden, M. L. Fogel, K. T. Howard, C. D. Herd, and L. R. Nittler(2012) The Provenances of Asteroids and Their Contributions to the Volatile Inventories of the Terrestrial Planets. Science 10 August 2012: 337 (6095), 721-723.

R. R. Becker and Mark A. Stahmann (1953) Polypeptide formation by reaction of N-Carboxyamino acid anhydrides in buffered aqueous solutions. In The Journal of Biological Chemistry October 1, 1953, 204, 737-744. [Cited 2010 Sep 12]. Available from: http://www.jbc.org/content/204/2/737.full.pdf+html

Jean-Michel Claverie, Chantal Abergel (2009) Mimivirus and its Virophage. In Annual Review in Genetics.43: pp. 49-66, 2009.

Jean-Michel Claverie, Chantal Abergel (2010) Mimivirus: the emerging paradox of quasi-autonomous viruses. In Trends in Genetics: pp. 431-437, August 7, 2010.

Bryan R. Cullen (2003) Nuclear RNA export. Journal of Cell Science 116, pp. 587-597. Available from: http://jcs.biologists.org/content/116/4/587.full.pdf

Seth A. Darst, Natacha Opalka, Pablo Chacon, Andrey Polyakov, Catherine Richter, Gongyi Zhang,and Willy Wriggers (2002) Conformational flexibility of bacterial RNA polymerase .Proceedings of the National Academy of Sciences of the United States of America Vol. 99, No. 7 (Apr. 2, 2002), pp. 4296-4301 Available from: http://www.pnas.org/cgi/doi/10.1073/pnas.052054099

Geoffrey S. Diemer and Kenneth M. Stedman (2012) A novel virus genome discovered in an extreme environment suggests recombination between unrelated groups of RNA and DNA viruses. In Biology Direct 2012, 7:13. Available from:

http://www.ncbi.nlm.nih.gov/pmc/articles/PMC3372434/pdf/1745-6150-7-13.pdf

C. Guerrier-Takada, K. Gardiner, T. Marsh, N. Pace, S. Altman (1983) The RNA moiety of ribonuclease P is the catalytic subunit of the enzyme. Cell. 1983 Dec;35(3 Pt 2)849-57.

Jaroslav Flegr (2009) A Possible Role of Intracellular Isoelectric Focusing in the Evolution of Eukaryotic Cells and Multicellular Organisms. Journal of Molecular Evolution November 2009, Volume 69, Issue 5, pp. 444-451. Available from: http://link.springer.com/article/10.1007%2Fs00239-009-9269-7/fulltext.html

Ricardo Flores, María-Eugenia Gas, Diego Molina-Serrano, María-Ángeles Nohales, Alberto Carbonell, Selma Gago, Marcos De la Peña, and José-Antonio Daròs (2009) Viroid Replication: Rolling-Circles, Enzymes and Ribozymes. Viruses. 2009 Sep; 1(2): 317–334. Published online 2009 Sep 14. doi: 10.3390/v1020317 PMCID: PMC3185496. Available from: http://www.ncbi.nlm.nih.gov/pmc/articles/PMC3185496/pdf/viruses-01-00317.pdf

Patrick Forterre, Jonathan Filée, and Hannu Myllykallio (2003) Origin and evolution of DNA and DNA replication machineries. In "The Genetic Code and the Origin of Life" L. Ribas, ed. Landes Bioscience. Available from:http://www.ncbi.nlm.nih.gov/books/NBK6360/

P. Jonathan, G. Butler, and Aaron Klug (1978) The Assembly of a Virus. In Scientific American November 1978. Republished in Scientific American Book Molecules to Living Cells. Publisher W. H. Freeman and Company, San Francisco 1980. Available from: http://www.med.unc.edu/biochem/carterlab/files/bioc-655/Butler_Klug.pdf

Eugene V Koonin, Tatiana G Senkevich, and Valerian V Dolja (2006) The ancient Virus World and evolution of cells. In Biology Direct 2006, 1:29. Available from: http://www.ncbi.nlm.nih.gov/pmc/articles/PMC1594570/pdf/1745-6150-1-29.pdf

Z. Martins, C. M. O'D. Alexander, G. E. Orzechowska, M. L. Fogel, P. Ehrenfreund (2007) Indigenous amino acids in primitive CR meteorites. In Meteoritics and Planetary Science December 2007. Available from: http://arxiv.org/ftp/arxiv/papers/0803/0803.0743.pdf

Hiroko Nashimoto and Masayasu Nomura (1970) Structure and Function of Bacterial Ribosomes, XI. Dependence of 50S Ribosomal Assembly on Simultaneous Assembly of 30S Subunits. In Proceedings of the National Academy of Sciences Vol. 67, No. 3, pp. 1440-1447, November 1970. Available from: http://www.pnas.org/content/67/3/1440.full.pdf

Matthew A. Pasek and Dante S. Lauretta (2005) Aqueous Corrosion of Phosphide Minerals from Iron Meteorites: A Highly Reactive Source of Prebiotic Phosphorus on the Surface of the Early Earth Astrobiology. 2005, 5(4): 515-535.

Lubert Strayer (1988) Biochemistry. New York: W. H. Freeman and Company. Book.

PHOTO AND ILLUSTRATION CREDITS

All photos and illustrations used to demonstrate cellular functions and viruses are public domain. Here I provide the required credits to the institutions that provided them:

Eduardo Trisapient Hernandez

Figures: 7, 8.

Maria Fernanda Hernandez

Figure: 20.

Darryl Leja created for the

National Human Genome Research Institute;

Figures: 2, 4, 6, 10, 11, 12, 13, 15, 17, 18, 19.

Jürgen Martens created for wikimedia.org;

Figures: 3, 5.

Dr. Fred Murphy and Sylvia Whitfield for the

Center for Disease Control and Prevention;

Figure: 15,16.

National Oceanic and Atmospheric Administration;

Figure: 1.

EMERGENT CHEMICAL EVOLUTION

COVER DESIGNED

By

The Author:

Eduardo Trisapient Hernandez

Using Gimp 2.8.6 on Ubuntu 12.04

And

Three Public Domain Pictures

Two from Darryl Leja for the National Human Genome Research Institute

And One from Виталий Смолыгин

http://www.publicdomainpictures.net/view-image.php?image=33832&picture=zoom-to-the-planet-and-the-moon&large=1